CONTENTS

CONTENTS

LIST OF SCENARIO BOXES

LIST OF TEXT BOXES

LIST OF FIGURES

LIST OF FIGURES

LIST OF TABLES

The **European Science Foundation** is an association of its 59 member research councils and academies in 21 countries. The ESF brings European scientists together to work on topics of common concern, to co-ordinate the use of expensive facilities, and to discover and define new endeavours that will benefit from a co-operative approach.

The scientific work sponsored by ESF includes basic research in the natural sciences, the medical and biosciences, the humanities and the social sciences.

The ESF links scholarship and research supported by its members and adds value by co-operation across national frontiers. Through its function as a co-ordinator, and also by holding workshops and conferences and by enabling researchers to visit and study in laboratories throughout Europe, the ESF works for the advancement of European science.

This volume arises from the work of the ESF Network for European Communications and Transport Activities Research (NECTAR).

Further information on ESF activities can be obtained from:

European Science Foundation
1 quai Lezay-Marnésia
67080 Strasbourg Cedex
France

Tel: 88 76 71 00
Fax: 88 38 05 32

PREFACE

This book has been written as part of the work of the *Europe 2020* group of the Network for European Communications and Transport Activities Research (NECTAR) organized by the European Science Foundation (ESF).

NECTAR was launched in 1986 in order to improve channels of communication and the coordination of research efforts in the fields of transport and communications between researchers and research institutions in Europe. Its first main task was a systematic evaluation of the state of the art of current research in transport and communications in Europe through a survey involving some 500 experts in the nineteen ESF member countries. In a second phase NECTAR has focused on four core research areas: (1) Barriers to Communication, (2) Europe 2020: Long-term Scenarios of Transport and Communications in Europe, (3) Supply and Demand Behaviour in Transport and Communications and (4) Transport and Communications Policy.

The *Europe 2020* group has looked into the long-term implications of transport and communications in Europe taking into account foreseeable changes in technology, industrial structures and social attitudes. Its work has resulted in three books: *Transport, Communications and Spatial Organization in the Europe of the Future*, edited by G. A. Giannopoulos, A. Gillespie and S. Wandel, *Logistics Platforms and City Logistics*, edited by C. Ruijgrok, B. Jansen, A. McKinnon and S. Wandel and this book, *The Geography of Europe's Futures*. The latter is the broadest in scope of the three. It sets out to consider the future of transport and communications in the context of long-range trends in the fields of population, life-styles, economy, environment, regional development and urban and rural form.

The book is a collaborative effort in every sense of the word. It is very much the product of a mutual learning process involving three individuals from different professional and national cultures. Throughout its two-year gestation period it has also been shaped and moulded at every stage by the comments and suggestions made by colleagues in the *Europe 2020* group. It owes a great debt to the sixty individual experts who unselfishly contributed the wealth of ideas and insights which enriched the overall discussion in the book. The authors wish to express their gratitude to all these colleagues and individuals whose names are listed at the end of the book. Finally, the authors would like to express their appreciation to Dr Iain Stevenson of Belhaven Press whose helpful and constructive comments on the work have played an important part in shaping the final product.

The contributions of several people at Dortmund and Sheffield also need to be mentioned here. At Dortmund University, student research assistants Simone Strähle and Seungil Lee spent many hours in collecting background material for the seed scenarios. Seungil Lee has developed tremendous skills in handling a drawing programme during the work on the book; he is responsible for most of the diagrams and maps. At Sheffield, Christine Hobson, Glenn Wilkins and Melanie Wood helped prepare the text for final publication.

Ian Masser
Ove Svidén
Michael Wegener
October 1991

ACKNOWLEDGEMENTS

The authors are grateful to the following copyright holders for permission to reproduce material: Yale University Press (Text box 2.1), Routledge (Text box 2.2), the estate of the late Sonia Brownell Orwell and Martin Secker and Warburg (Text box 2.3), Duckworth (Text box 2.4), The Royal Society of Arts (Text box 4.1), Klaus Kunzmann (Text box 8.1), The World Futures Society (Text box 11.1), the International Institute of Applied Systems Analysis (Fig 6.1), David Keeble (Fig 7.3), Réclus (Fig 7.4), Eric Verroen and Giselbertus Jansen (Fig 8.5), the European Conference of Ministers of Transport (Fig 9.2), Dieter Lapple (Fig 9.4), Fleet/Drive (Fig 9.5), Mens en Ruinste (Fig 10.4) and the Commission of the European Communities (Fig 11.1).

Ian Masser
Ove Svidén
Michael Wegener
February 1992

Part I
Exploring the future

The aim of this book is to stimulate a more informed debate about the future spatial organization of Europe. By spatial organization we mean the allocation of human activities in space and the interactions between them. Human activities, as production, distribution, housing and leisure, have specific space requirements and at the same time are connected by a variety of interactions such as movements of people or goods or flows of information. Because of the functional relationship between location and interaction, transport and communications are fundamental for the organization of space. For this reason a book on the future geography of Europe is one about the future of *transport and communications in a spatial context*. However, as its title suggests, the book does not address one single 'future' only. Underlying the choice of title is the belief that the future is not predetermined. It can be shaped so long as there is the will and the determination to shape it. For this reason the authors have not attempted to make 'accurate' predictions about the future but rather to present a number of alternative scenarios or pictures of possible *futures*.

One thing we can be sure about is that all the scenarios presented in the book will be wrong in some measure. However, the value of the book is not to tell us what *will* happen but to help us think constructively about the choices that lie before us and the decisions that need to be made to increase the chances of more desirable options and decrease the chances of less desirable options. In other words, we can play an important role in shaping the geography of Europe's futures.

At the outset it must be admitted that there is a strong element of wishful thinking in any exercise of this kind. However, it is not the intention of the authors to present pictures of an idealized future state or Utopia or its antithesis, Dystopia. The scenarios that have developed in the book reflect different combinations of assumptions about changes in both trends and values between now and 2020. The common feature of these scenarios, then, is that they represent realizable rather than idealized futures relating to our working and living environment in Europe as a whole.

The book challenges the reader to explore the properties of these scenarios with particular reference to the likelihood that they will be predominant in 2020 and their desirability in relation to what the reader would like to see happen by that date. To help the reader the authors have called upon the services of more than 60 experts from all West European countries. These experts have commented extensively on the likelihood of the different scenarios coming into being and they have also given their own views as to which scenarios they prefer for their own countries. Their views add to the richness of the debate as well as give an indication of the current climate of opinion amongst some key participants in opinion forming and decision-making in the transport and communications field.

The first two chapters set the scene for the detailed discussion that is contained in the rest of the book. Chapter 1 sets out the reasons for the choice of the scenario method and explains some of the key assumptions that underlie the whole work. Chapter 2 considers the broader context within which works of this kind are set. It draws upon the rich heritage of past futures studies to highlight a number of key issues relating to the form, content and process of future futures studies.

CHAPTER 1

INTRODUCTION

What will Europe look like in, say the year 2020? One thing we can be sure about is that it will be different from the present. But different in what ways? Opinions vary as to the likelihood and the desirability of various developments. This is because the future is not known. It will be created. Trends can be analysed and projections can be made about the future but there are often marked breaks from past trends, and policies which have been accepted for decades can suddenly lose their power as values and attitudes change.

The future geography of Europe is therefore the subject of debate:

— What is most likely to happen?
— Is this the most desirable choice?
— If not, what can be done?

This book is a contribution to the debate about Europe's futures. It explores current trends which affect our working and living environment in relation to three contrasting scenarios of Europe in 2020. It seeks to stimulate debate by highlighting the policy choices that are open to us as Europeans to shape our future. In particular, the book explores the impacts of current developments in transport and communications in the context of socio-economic, political and environmental change. The questions it addresses include:

— Will the creation of the Single European Market lead to a further concentration of activities in the core belt stretching from London to Milan?
— What will be the socio-economic impacts if the use of the automobile is drastically curtailed for environmental reasons?
— What will be the impact on urban and rural form of the decentralization of economic activities from large urban centres?
— Will the introduction of fibre optics technology in communications reduce regional disparities within Europe by 2020?

Why scenarios?

Scenarios are descriptions of future developments based on explicit assumptions. As a method for exploring the future scenarios are superior to more rigorous forecasting methods such as statistical extrapolation or mathematical models if the number of factors to be considered and the degree of uncertainty about the future are high. This clearly applies in the case of transport and communications. Transport and communications are closely interrelated with almost all aspects of human life. They are linked to social and economic developments, are influenced by technological innovations and are subject to numerous political and institutional constraints.

In the face of this complexity scenarios are perhaps the only method to identify corridors of relevant and feasible futures within a universe of possible ones. Moreover, scenarios have, in relative terms, only moderate data requirements, permit the incorporation of qualitative expert judgement and, in conjunction with appropriate techniques such as the Delphi method, facilitate the process of converging initially different expert views towards one or possibly a few dominant opinions. In addition, scenario writing as a group exercise has the potential of generating awareness of factors and impacts which may not have been identified through formal forecasting methods.

The year 2020 has been chosen as the forecasting horizon. This may seem rather a long time if one considers the speed of change in the socio-economic and technological context of transport and communications. However, transport and communications infrastructure, because of the heavy investment involved, changes only slowly and the introduction of a fundamentally new transport or communications technology such as high-speed trains or ISDN (Integrated Services Digital Network) may even require decades to complete. Conversely, if one looks at the impacts of new transport or communication systems on, say, land use, or the location of households and firms, these changes become effective only with considerable time-lags and even more time is involved before the changes in travel behaviour induced by these land-use changes are felt in the transport system. Political changes are sometimes even slower, as the history of the adaptation of standards between national railway systems in Europe or the slow diffusion of pollution control for cars in some European countries demonstrates. For all these reasons it was felt to be necessary to study the future of transport and communications in Europe within a thirty-year framework.

Three scenarios

Before any scenarios can be developed it is necessary to ask which kind of Europe is envisaged. After the rapid changes in eastern Europe over the last year and following the Gulf War no question could be more difficult to answer. However, it is necessary to fix ideas. Therefore without being too specific, a few general assumptions are presented:

— *In 2020 'Europe' will be larger than the current EC.* Most likely some or all of the eastern European countries and the countries now forming EFTA will have joined the European federation in some form. Altogether the European federation will encompass between 400 and 500 million people, more than twice as many as the USA and the Pacific countries.
— *In 2020 there will be a European government.* Most likely Europe will be a federation of more or less autonomous countries each with its own legislation, jurisdiction and government. Nevertheless, there will be a European president, a European cabinet and a European parliament with significant powers over member states where European matters are concerned. International Trade and Industry, Research and Technology, Environment and Transport and Communications will be the most prominent European ministries because the need for integrated European policy-making is most obvious in these fields.
— *In 2020 there will be peace in Europe.* Of course this is more a hope than a scientific hypothesis but it is a necessary one if one is to make any predictions about the future. In short, it is assumed that between now and the year 2020 there will be no major political economic crises, climatic or nuclear catastrophes, civil wars or military aggressions that will substantially disrupt the peaceful process of European integration.

Beyond the assumptions stated above, everything else is left open. No assumptions are made about the events that will shape policy-making in local, regional European governments. However, by identifying a set of different political directions or paradigms a domain for the future evolution of transport and communications in Europe is opened up. In this top-down approach the points of departure are three major directions or global scenarios. Tentatively they are associated with the key words *growth*, *equity* and *environment*.

— *The growth scenario* shows the most likely development of transport and communications in Europe if all policies emphasize economic growth as the primary objective. This would most probably also be a high-tech and market economy scenario with as little state intervention as possible. This scenario might be associated with the political ideals of many current conservative governments in Europe.

— *The equity scenario* shows the impacts of policies that primarily try to reduce inequalities in society both in terms of social and spatial disparities. Where these policies are in conflict with economic growth, considerations of equal access and equity are given priority. This scenario might be associated with the typical policy-making of social democratic governments.

— *The environment scenario* emphasizes the quality of life and environmental aspects. There will be a restrained use of technology and some control over economic activity. Where economic activities are in conflict with environmental objectives a lower rate of economic growth will be accepted. This scenario might be associated with the views of the green parties throughout Europe.

The relationship between the three paradigms or political directions is illustrated by the triangle shown in Figure 1.1. Each of its corners presents one of the paradigms: growth, equity or environment. The triangle area represents the domain of possible changes from the present condition. The line starting from the centre is the trajectory from our present state to the distant future of Europe 2020: it may bend in response to technical breakthroughs, new organizational patterns or political decisions.

The component scenarios

The great advantage of the scenario approach is that we can explore alternative visions of the future in as much detail as we wish. These visions combine elements from all the main problem areas that impinge on the geography of Europe's futures. In practice, however, it is generally preferable to deal with one issue at a time in some depth rather than trying to grapple with too many questions at once. Consequently, nine problem areas or fields have been identified for separate treatment prior to the evaluation of the comprehensive scenarios. The main features of these component scenarios are described in Table 1.1.

The nine component scenarios can be grouped into three broad clusters of topics. The first of these deals with the social and economic context within which future changes in transport and communications will take place in Europe. It considers matters relating to changes in *population, life-styles* and *economy*. The second cluster is concerned with the spatial dimension of transport and communications. It examines issues with changes in the *Environment, regional development*

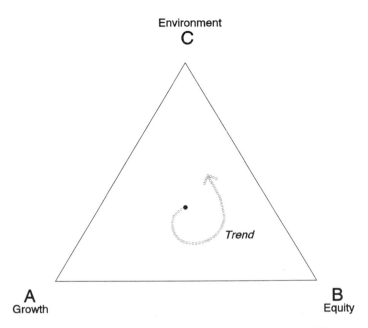

Figure 1.1 The underlying paradigms. Three scenarios based on the different political directions or paradigms associated with growth, equity and the environment are used to open up the debate on the geography of Europe's futures. The present situation is indicated at the centre of the triangle. The line starting from the centre is one possible trajectory from our present state to that of Europe in 2020.

and *urban and rural form*. The last cluster explores developments in transport and communications themselves. It evaluates the impact to the changes that are currently taking place with respect to *goods transport, passenger transport* and *communications*.

Each component scenario describes the likely development in its field in relation to one of the three global scenarios: that is to say, each component scenario is associated with one global scenario and there are three component scenarios for each field (Table 1.2). The component scenarios for each field differ according to the assumptions that are made about the speed or even direction of change, developments in other fields (other component scenarios), or the global scenario with which they are associated.

However, there are no separate component scenarios for different countries: all scenarios have been written from a European perspective. They describe developments in Europe as a whole taking account of different trends in different countries wherever necessary. Thus different speeds of development in different types of countries can be addressed.

Table 1.1 Key features of the component scenarios

Field	Definition
Population	Changes in fertility, mortality and migration and their impacts on the overall age structure of the population
Life-styles	Changes in household size and composition, labour force participation and activity patterns
Economy	Including economic structural change to a post-industrial society and its impacts on industrial organization and reorganization
Environment	The use of energy, air and land resources and factors governing human well-being, safety and protection from noise and physical disturbance
Regional development	Growth and decline processes in central and peripheral regions of Europe and the spatial disparities resulting from them
Urban and rural form	Changes in the internal structure of regions and the relationship between cities and their hinterlands
Goods transport	Changes in the volume and directions of bulk cargo, raw materials, energy food and industry products movements
Passenger transport	Changes in personal mobility, the volume and intensity of trip movements, and the use of different modes of travel
Communications	The emergence of the information society, restructuring of telecommunications and information handling

Note: The nine component scenarios that will be discussed in Part II of this book represent the main problem areas impinging on urban and regional development and transport and communications in Europe.

Of course, the real world is not so well ordered as Table 1.2 suggests. First of all, the actual development is not likely to follow one of the three global scenarios in its pure form. The Europe of the year 2020 will be a federation in which different political directions are likely to coexist in different countries and even within the countries there may be different styles of policy-making dominant in different regions or at different times. Second, the nine fields are not nearly as mutually exclusive as the neat rows of the matrix imply. In the real world they overlap and are linked by an intricate cobweb of mutual interdependencies: population and economy interact on the labour market, consume environmental resources, determine regional development and urban and rural form and generate flows

Table 1.2 Scenarios and component scenarios

	Scenarios		
	A	B	C
	Growth	Equity	Environment
Component scenarios			
1 Population	A1	B1	C1
2 Life-styles	A2	B2	C2
3 Economy	A3	B3	C3
4 Environment	A4	B4	C4
5 Regional development	A5	B5	C5
6 Urban and rural form	A6	B6	C6
7 Goods transport	A7	B7	C7
8 Passenger transport	A8	B8	C8
9 Communications	A9	B9	C9

Note: The three scenarios and their component scenarios can be arranged in matrix form. Each row represents one of the fields while each column represents a seed scenario. Each cell (A1, B1, C1, etc.) represents a component scenario.

of goods, passengers and information which in turn co-determine the process of spatial development, affect the environment and give rise to new mobility patterns and life-styles. All these interdependencies have to be kept in mind when dealing with the component scenarios.

Their interpretation

Scenarios such as those described above are only of limited value in themselves unless they can be subjected to critical evaluation. The essential problem facing most scenario writers is to find experts who are capable of evaluating their findings and putting forward their own interpretations of events. In this case, however, the interpretation of the scenarios presented no problem because the authors have been able to draw upon the unique body of collective knowledge about past and anticipated trends affecting transport, communications and mobility in Europe as a whole that has been built up by more than sixty scholars from nineteen European countries who are participating in the Network for European Transport and Communication Activities Research (NECTAR) funded by the European Science Foundation.

The knowledge and experience of these scholars and their associates has been tapped in the detailed evaluation of the three scenarios for each of the nine fields. These experts form a Delphi panel which was also asked to indicate which option they regarded as the most likely scenario and which they would like to see in their own countries. In this way the interpretation of the findings of the scenarios by this broad group of experts opens up the discussion and points to the choices facing European decision-makers.

The Delphi panel consists of scholars drawn from a wide range of academic disciplines and different types of practical experience. Figure 1.2 shows that it includes engineers, economists, urban planners and geographers of varying ages and experience from all parts of Europe. This experience ranges from urban modelling and analysis to practical policy advice to governments regarding transportation and communications projects. However, they have one thing in common as a result of their involvement in NECTAR. That is, they are all actively engaged in research at the European as against the national level.

The rest of the book

The rest of this book consists of twelve chapters which develop the themes set out above. The next chapter describes in more detail the background to scenario writing and discusses some of the findings of earlier

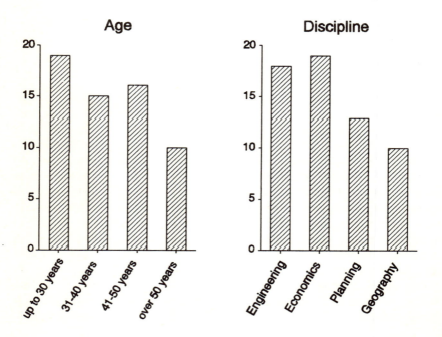

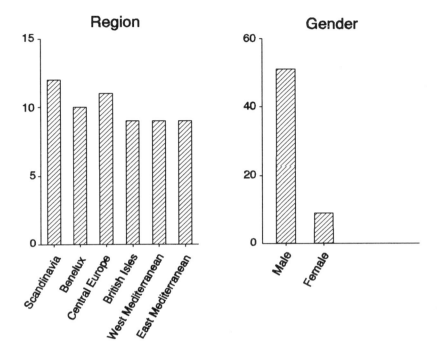

Figure 1.2 Profile of the experts. The members of the Network for European Transport and Communications Activities Research (NECTAR) who have interpreted the scenarios have a wide range of academic and practical experience. The participants are spread throughout the 20 to 60 age-group, and a broad spectrum of disciplines is represented including engineering, economics, geography and planning. They are drawn from all the main subregions of Europe.

studies of European futures. Chapter 3 to 11 form the heart of the book. These chapters describe the geography of Europe's futures in relation to each of the nine components described above. To assist in the presentation, a common format is used in each of these chapters. The last part of the book is divided into two chapters. The first of these synthesizes the findings of the evaluation while the second highlights some of the key policy choices facing European decision-makers in the light of these findings.

Further reading

The first phase of the Network for European Communications and Transport Activities Research (NECTAR) involved a comprehensive survey of past and current trends in the socio-economic and technological contexts of transport, communications and mobility; in travel and communications behaviour; and in the political and institutional framework underlying policy formulation in the nineteen European Science Foundation countries. This work draws heavily on the findings of this survey which are discussed in a book by Nijkamp *et al.* (1990).

CHAPTER 2

ANTECEDENTS

The opening chapter set the scene for the rest of this book. This chapter examines in greater detail the main issues relating to the content, form and process of futures related studies that must be taken account of in the development of scenarios for exploring the geography of Europe's futures.

The heritage of futures studies

The scenarios developed in Part II of this book draw upon a substantial heritage of futures related to speculations and conjectures that goes back at least 2,000 years to Plato's description of his ideal republic. The studies take a variety of forms from polemic to science fiction. In the last thirty years futures research has emerged as a scholarly field in its own right and new methodologies have been developed for the task. The growth of futures related research reflects not only our insatiable curiosity about what might be or what could be but also a growing recognition of the value of studies of this kind for strategic decision-making in the fields of new product development, technological innovation, defence studies and economic and social change.

Some of the most important features of this heritage can be seen in the four extracts contained in Text Boxes 2.1 to 2.4. The first of these extracts is taken from Book II of Thomas More's *Utopia*. Published in 1518, More's *Utopia* occupies a unique position in the history of future studies. It epitomizes the ideals of humanism that characterize the Renaissance in Europe. It is an apocalyptic vision of the best earthly state possible—Utopia.

What distinguishes *Utopia* from other writing of its time is that it paints a dramatic picture of what was expounded theoretically by other scholars and writers. In this respect the work can be regarded as a precursor of what has subsequently come to be known as the scenario method.

UTOPIA

The island of the Utopians extends in the centre (where it is broadest) for two hundred miles and is not much narrower for the greater part of the island.

The island contains fifty-four city-states, all spacious and magnificent, identical in language, traditions, customs and laws.

In the senate at Amaurotum (to which, as I said before, three are sent annually from every city), they first determine what commodity is in plenty in each particular place and again where on the island the crops have been meagre. They at once fill up the scarcity of one place by the surplus of another. This service they perform without payment, receiving nothing in return from those to whom they give. Those who have given out of their stock to any particular city without requiring any return from it receive what they lack from another to which they have given nothing. Thus the whole island is like a single family.

But when they have made sufficient provision for themselves (which they do not consider complete until they have provided for two years to come, on account of the next year's uncertain crop), then they export into other countries, out of their surplus, a great quantity of grain, honey, wool, linen, timber, scarlet and purple dyestuffs, hides, wax, tallow, leather, as well as livestock. Of all these commodities they bestow the seventh part on the poor of the district and sell the rest at a moderate price.

The intervals between the hours of work, sleep and food are left to every man's discretion, not to waste in revelry or idleness, but to devote the time free from work to some other occupation according to taste. These periods are commonly devoted to intellectual pursuits. For it is their custom that public lectures are daily delivered in the hours before daybreak.

Text Box 2.1 Extracts from Thomas More's *Utopia* (1518) (Surtz, 1964). *Utopia* embodies the spirit of the Renaissance and the ideals of humanism. This description of life in the best of all worlds is a precursor of the scenario method.

As can be seen from Text Box 2.1, More's Utopia bears a strong resemblance to Britain in terms of its size, and its capital city Amaurotum is modelled on London. The way of life that is described is one of voluntary simplicity with great importance attached to self improvement through education. As More points out, the whole island is like a single family, and great care is taken to ensure that the needs of the less well-off are taken care of in the interests of the community as a whole.

In such a world governed by tolerance and self-restraint there is no room for individual greed and ambition. More's Utopia is both a stable and a static society. Nor is there any pretext for evading work. 'There is no workshop, no alehouse, no brothel anywhere, nor opportunity for corruption, no lurking hole, no secret meeting place.'

William Morris described *News from Nowhere* as a Utopian romance. It was published in instalments between January and October 1890. Although it was written nearly four centuries after *Utopia* it has many features in common with its famous predecessor. Morris also depicts, in very vivid terms, what life might be like in an idealized society sometime in the future. As in Utopia this society is both stable and static in nature. Both More and Morris are critical of established power. In More's case his critique of the growth of absolutism in the hands of kings such as Henry VIII is implicit rather than explicit. In Morris's case, however, the critique of the evils of capitalist society is central to the work. As a follower of Karl Marx, William Morris sought to paint a picture of a classless Utopia of the twenty-first century.

The extracts from *News for Nowhere* that are contained in Text Box 2.2 show how life in London has changed since the capitalist system was overthrown. The business premises which Morris regards as little more than gambling dens have been taken over by the former slum dwellers who celebrate The Clearing of Misery every year. Outside central London, Britain has become once more a garden of healthy and happy rural communities.

There is a strong element of nostalgia in Morris' work. The spirit of fourteenth-century England pervades his twenty-first-century Utopia. Capitalism with its big industrial cities and its exploitation of the resources of the countryside must be replaced by a truly Communist society where power is thoroughly decentralized. In such a society, symbols of power such as the Houses of Parliament become 'a sort of subsidiary market and storage place for manure'.

George Orwell's bleak description of London in *Nineteen Eighty-Four* is in stark contrast to the humanist and socialist idealism that characterizes *Utopia* and *News from Nowhere*. Orwell wrote *Nineteen Eighty-Four* in the years following the defeat of Fascist Germany in the Second World War and the nuclear holocausts of Hiroshima and Nagasaki. His novel is an awful warning of what life might be like in a future totalitarian state. Such a dystopia where everything is as depressingly wretched as possible represents the antithesis to the Utopian dream.

The extracts from the book that are contained in Text Box 2.3 depict Orwell's hero Winston Smith in his seventh-floor flat in Victory Mansions looking out on an impoverished London ravaged by continuous wars. In

NEWS FROM NOWHERE

Once a year, on May-day, we hold a solemn feast in those easterly communes of London to commemorate The Clearing of Misery, as it is called. On that day we have music and dancing, and merry games and happy feasting on the site of some of the worst of the old slums, the traditional memory of which we have kept. On that occasion the custom is for the prettiest girls to sing some of the old revolutionary songs, and those which were the groans of the discontent, once so hopeless, on the very spots where those terrible crimes of class-murder were committed day by day for so many years.

Our forefathers, in the first clearing of the slums were not in a hurry to pull down the houses in what was called at the end of the nineteenth century the business quarter of the town . . . You see, these houses, though they stood hideously thick on the ground, were roomy and fairly solid in building, and clean, because they were not used for living in, but as mere gambling booths; so the poor people from the cleared slums took them for lodgings and dwelt there, till the folk of those days had time to think of something better for them. So used to living thicker on the ground there than in most places; therefore it remains the most populous part of London; or perhaps of all these islands . . . However, this crowding, if it may be called so, does not go further than a street called Aldgate, a name which perhaps you may have heard of. Beyond that the houses are scattered wide about the meadows there, which are very beautiful, especially when you get on to the lovely river Lea.

This is how we stand. England was once a country of clearings amongst the woods and wastes, with a few towns interspaced, which were fortresses for the feudal army, markets for the folk, gathering places for the craftsmen. It then became a country of huge and foul workshops and fouler gambling-dens, surrounded by an ill-kept, poverty-stricken farm pillaged by the masters of the workshops. It is now a garden, where nothing is wasted and nothing is spoilt, with the necessary dwellings, sheds and workshops scattered up and down the country, all trim and neat and pretty.

Text Box 2.2 Extracts from William Morris' News from Nowhere (1890) (Redmond, 1970). This Utopian romance depicts life in the classless socialist society which came into being after the overthrow of the capitalist system in the twentieth century.

NINETEEN EIGHTY-FOUR

Winston kept his back turned to the telescreen. It was safer; though, as he well knew, even a back can be revealing. A kilometre away the Ministry of Truth, his place of work, towered vast and white above the grimy landscape. This, he thought with a sort of vague distaste— this was London, chief city of Airstrip One, itself the third most populous of the provinces of Oceania. He tried to squeeze out some childhood memory that should tell him whether London had always been quite like this. Were there always these vistas of rotting nineteenth-century houses, their sides shored up with baulks of timber, their windows patched with cardboard and their roofs with corrugated iron, their crazy garden walls sagging in all directions? And the bombed sites where the plaster dust swirled in the air and the willow-herb straggled over the heaps of rubble; and the places where the bombs had cleared a large patch and there had sprung up sordid colonies of wooden dwellings like chicken-houses? But it was no use, he could not remember: nothing remained of his childhood except a series of bright-lit tableaux, occurring against no background and mostly unintelligible.

The Ministry of Truth—Minitrue, in Newspeak—was startlingly different from any other object in sight. It was an enormous pyramidal structure of glittering white concrete, soaring up, terrace after terrace, three hundred metres into the air . . . Scattered about London there were just three other buildings of similar appearance and size. So completely did they dwarf the surrounding architecture that from the roof of Victory Mansions you could see all four of them simultaneously. They were the homes of the four Ministries between which the entire apparatus of government was divided. The Ministry of Truth, which concerned itself with news, entertainment, education and the fine arts. The Ministry of Peace, which concerned itself with war. The Ministry of Love, which maintained law and order. And the Ministry of Plenty, which was responsible for economic affairs. Their names, in Newspeak: Minitrue, Minipax, Miniluv and Miniplenty.

Text Box 2.3 Extracts from George Orwell's *Nineteen Eighty-Four* (1949) (Crick, 1984). This work presents the antithesis of the idealized worlds described by More and Morris. Orwell's dystopia is a chilling account of terror and violence in a totalitarian state.

his world all power lies in the hands of those who control the State, the Party. They have developed a wide range of procedures for moulding public opinion to their own ends and eliminating dissenters. Big Brother is always watching you in such a world and history is constantly being rewritten at the Ministry of Truth.

Orwell's novel describes the powerful mechanisms that have been used by both Fascist and Communist regimes to maintain them in power and sustain a class system which exploits the 'proles' without mercy. Under such circumstances objective truth has little meaning. In the words of the Party's slogans 'war is peace, freedom is slavery and ignorance is strength'.

The last of the four extracts is the only one which originated in a formal scenario writing exercise. Between 1967 and 1975 more than 200 experts from ten countries took part in the *Europe 2000* project which was sponsored by the European Cultural Foundation. This project made extensive use of the scenario method to explore a wide range of futures with respect to the international position of Europe relative to the world as a whole, the long-term consequences of changes in the economy and society and possible trends in urbanization and environment.

The findings of this project suggest that Europe in the year 2000 will be a transitional society between that which it was in being around 1970 when the research was carried out and a quite different one. By the year 2000 Europe will be well on the way towards a resource-saving society which will operate in an increasingly decentralized way.

Conservation and decentralization feature prominently in the extracts from the last chapter of the *Europe 2000* report that are contained in Text Box 2.4. In contrast to *News from Nowhere* and *Nineteen Eighty-Four*, the choice of location is not London but a rural area within easy reach of one of the major metropolitan centres. This does not imply that London will cease to exist. Big cities and industrial mass production will continue, but the most significant changes are likely to be in rural rather than urban life. The decentralization of power is manifested in an increasingly dispersed pattern of settlements. The significance that is attached to environment conservation is evident in the choice of the family's dwelling and the measures that have been devised to deal with the transport needs of the community.

The content, form and process of futures studies

The heritage of futures studies is a very rich one and the four examples discussed above do it scant justice. Nevertheless, they draw attention to a

EUROPE 2000

One typical European family of the year 2000—we can call them the Dumills or the Deuxmilles or the Zweitausends—live in a converted eighteenth-century farmhouse on the edge of a hill area between 70 and 150 kilometres from a major city; we can imagine them in the Peak District or the Pentlands, or the Ardennes, or the Eifel or the Monts de Morvan or the Sierra de Guadaramma. Built in an energy conscious age, this farmhouse has properties of insulation which make it very apposite to a new age of conservation.

The farm is one of a group forming a small rural hamlet. It is occupied by a number of families that moved into them after they were abandoned in the late 1950s during the great age of European agricultural depopulation. Lower down the valley are other such family groups, forming a loose cluster of about fifty nuclear families or about two hundred people. Together with other clusters and the nearest village they form a sufficiently large group to support a village primary school and community centre.

Such a dispersed rural pattern of life, it might be thought, must place big demands on resources for transportation. But those demands have been limited in a number of ways. First, because of the varied character of the rural population it is able to satisfy so many of its social and cultural needs locally. Secondly, the development of information technology has been so rapid that many needs are met without having to travel at all . . . Thirdly, because the age of expensive energy has created its own response in the form of energy conserving vehicles and organisational arrangements. To move about locally, most villages use several mopeds in which the motor is used only as a supplementary device. To move longer distances, they rely on a system of shared rides whereby anyone leaving the village is under an obligation to offer seats in his car, truck or van.

Text Box 2.4 Extracts from *Europe 2000* (Hall, 1976). The *Europe 2000* project made extensive use of the scenario method to explore alternative futures, drawing upon the knowledge of 200 experts in ten countries. The findings of the research suggest that Europe will be living in an increasingly decentralized resource saving society by the year 2000.

number of features relating to the content, form and process of futures studies which must be taken into account in the development of the scenarios in the next part of the book.

Content

One lesson that can be drawn from the four examples is the extent to which the underlying philosophical positions change only slowly over time. In Chapter 1 it was argued that these can be grouped into three broad categories representing the goals of growth, equity and environment respectively. The idea of progress which emerged from humanistic thinkers of the Renaissance such as Thomas More underlies the growth paradigm. Similarly the equity paradigm is central to the work of socialists such as William Morris while the environment paradigm features prominently in the *Europe 2000* scenario. In this way the examples highlight the extent to which the three paradigms in the next section reflect long-standing and very well developed intellectual traditions.

Another lesson that can be drawn from the four examples is that there is a strong polemical dimension in all future studies. Their purpose is not merely to inform but also to present opinions. Such studies differ from mere extrapolations of current trends into the future. This does not mean to say that these studies ignore such trends. What it implies is that these trends are interpreted in relation to different underlying philosophical positions. As a result, scenario writing is a creative activity which is fundamentally different from trend extrapolation even though it draws on the same materials.

A corollary of their polemical nature, however, is that many future studies have a conservative if not authoritarian undertone. Whether intended as positive counter-worlds to a negatively perceived present or as negative extrapolations of tendencies perceived as harmful, Utopian writing tends to project strongly regulated social systems. Plato's *Republic*, Campanella's *Sun City*, Bacon's *The New Atlantis* and More's *Utopia* are clearly authoritarian societies. So are Orwell's *Nineteen Eighty-Four* and Huxley's *Brave New World* in a negative sense. Even Utopian scenarios which were explicitly written *against* rules and restrictions perceived as obsolete such as Morris' *News from Nowhere* or Skinner's *Walden 2* are full of meticulously, though allegedly voluntary, observed conventions. Anarchy or rebellion are the exception in Utopian writing as they run counter to the intention of their authors to show a positive or negative model society.

Form

All four examples show that the great strength of the scenario method is its ability to stir the imagination by presenting vivid pictures of future states. These pictures challenge the reader to draw upon his or her own experience to judge whether such a state is desirable or to be deplored. It also forces him or her to think more critically about the present as well as the future and helps to identify areas of choice.

An interesting observation can be made about the use of space and time in future studies. Early Utopias were located in an imaginary part of the world so far away that the contemporary reader was unlikely to ever have the chance to verify the existence of the legendary state. This changed when in the seventeenth and eighteenth centuries even the remotest oceans and islands were surveyed and mapped. From then on Utopias moved to a distant future.

A particularly powerful tool in the hands of the scenario writer is the metaphor. The Clearing of Misery festival described by William Morris conjures up a host of images. Similarly Orwell's monumental ministry buildings symbolize the power of the totalitarian state.

From the examples it can be seen that the impact of the scenario method is greatest where it highlights differences between present and possible future states. Yet it must be recognized that, like the *Europe 2000* project, all future societies are in some measure transitional societies. Many features of these societies will resemble those of present societies as the rate of change in life-styles and urban development or even transport and communications is relatively slow.

Process

Each of the four examples discussed above is the product of a long process of gestation. The antecedents of More's *Utopia*, William Morris' *News from Nowhere* and Orwell's *Nineteen Eighty-Four* have been extensively debated by literary critics. In the case of the *Europe 2000* project, however, a systematic attempt was also made to draw upon the collective experiences of 200 experts from ten countries in the process of scenario development. The value of using the collective knowledge of a large number of experts in the development of scenarios is well documented in the futures research literature. An alternative to the technique used in the *Europe 2000* project is to invite experts to participate in the evaluation of previously constructed scenarios. This, as can be seen in the next part of this book, has considerable advantages from the standpoint of increasing the richness and diversity of the scenarios that are presented to the reader.

Further reading

Over the last quarter of a century there has been a marked expansion in the whole field of futures studies. The field has developed its own professional associations and generated scholarly journals such as *Futures* and *Technological Forecasting and Social Change*. The best introduction to this field is the *Handbook of Futures Research* edited by Jib Fowles (1978). A more specific discussion of scenario methods in the planning field can be found in Hirschhorn (1980).

Some scenario-based studies have generated widespread public debate throughout the world. Two of the best known of these are the Hudson Institute's study of *The Year 2000* (Kahn and Wiener, 1967), which presented a number of technical, economic and social trends and political scenarios, as well as some nightmare defence scenarios, and the Club of Rome's analysis of *The Limits to Growth* (Meadows *et al.*, 1972), which explored the environmental consequences of population growth and economic development. The broader issues raised by studies such as these are discussed in some detail in the volume edited by Freeman and Jahoda (1978).

Both the scenario and Delphi methods have been widely used in urban planning and transport planning. A good example of the use of the scenario method in urban and regional planning is Peter Hall's *London 2000* (1963) which has been recently updated and republished as *London 2001* (1989). An interesting example in the transport field which utilizes both the scenario and the Delphi methods is Svidén's (1988) study of future information systems for road transport. A useful overview of the Delphi method can be found in Masser and Foley's (1987) study of the long-term prospects for the Sheffield economy.

One indication of the extent of interest in the field of futures studies is the extent to which books on this topic have reached best-seller status. Good examples of these which contain some valuable insights on the issues discussed in Part II of this book are Alvin Toffler's *Future Shock* (1970) and *The Third Wave* (1980), John Naisbitt's *Megatrends* (1984) and its successor, *Megatrends 2000* (Naisbitt and Aburdene, 1990) and Charles Handy's *The Age of Unreason* (1989).

Part II
Scenarios

There are various ways of collecting the information for writing scenarios. One way is to study the literature, interview experts, or rely on one's own knowledge, theoretical insight, experience or imagination. In this case the scenario or scenarios are the final product. Another way is to use the scenario as a tool for soliciting information, ideas, agreement or criticism or suggestions for change from knowledgeable individuals. In that case the scenario or scenarios are written at the beginning of the exercise.

Such scenarios written *a priori* are called *seed scenarios* because they are used as seeds to stimulate a chain reaction of cross-fertilizing thoughts and associative ideas. Obviously, seed scenarios, though they may be prepared with the best effort and after consultation of all available sources of information, are liable to be sketchy and incomplete, sometimes even erroneous—if it were otherwise, they would be the final product.

The scenarios presented in the following chapters are seed scenarios. They were used in the exercise to introduce the respondents to the topic in question, to solicit confirmation or objections and to stimulate comments and additional suggestions.

As explained in Chapter 1, nine problem areas relevant for the future development of transport and communications in Europe were identified (see Table 1.1). Each of the following nine chapters deals with one component area or field. To assist in the presentation of the material, a common format is used in each of the chapters:

— Each chapter begins with a short section containing a definition of the field. This highlights the most important relationships between the problem area and those considered in the other eight chapters.
— In order to facilitate the assessment of the contents and likelihood of

the component scenarios, in a second section common background information is provided including information on past trends, present condition, most likely future trends, opportunities and constraints and policies. The 'most likely future trends' represent in fact a 'trend scenario', i.e. an extrapolation of current social, economic, technological and political tendencies. It shows the most likely development of transport and communications in Europe *if no major disruptions or breaks in the socio-economic, technological and political context occur.* This implies of course that policies already 'in the pipeline', such as the completion of the Single European Market, are taken into account. The component scenarios use this background information as their common point of departure.

— As explained in Chapter 1, for each of the nine fields three scenarios were developed: the *growth scenario* (A), the *equity scenario* (B) and the *environment scenario* (C). In the third main section of each chapter, these three component scenarios are presented in their original form and elaborated and expanded in the light of the comments and suggestions of the respondents.

— The fourth section presents the views of the experts from the NECTAR group panel who evaluated the three alternative scenarios. Particular attention is given to the variations on the scenarios that were suggested by the experts and their assessment of the likelihood and desirability of them coming into being.

— The final section is left to the authors' evaluation of the results: Which scenario is most likely? Will this be uniform all over Europe? Which aspects of it are desirable or undesirable? What are the choices for policymakers? What needs to be done to implement them?

In this way each of the nine chapters that make up this part of the book describes the main features of current trends in a key problem area, presents three alternative scenarios of Europe 2020, and discusses the findings of the expert evaluation of these scenarios. Once this has been achieved for all the nine problem areas, the way is clear for the overall assessment of the scenarios that can be found in Part III of this book.

Finally, a note to readers. To preserve the original flavour of the exercise, the trend information and seed scenarios are reproduced here with only minimal editing as they were presented to the respondents. Space constraints and time constraints on the side of the respondents dictated the format of presentation, which frequently had to resort to metaphors and symbols where a comprehensive treatment would have been desirable. Readers are asked to add their own imagination.

CHAPTER 3

POPULATION

Definition

Future changes in population in Europe will have a marked effect on most of the other seed scenarios. Changes in the overall size of Europe's population will be reflected in increased (or reduced) pressures on environmental resources (see Chapter 6) and urban land uses (see Chapter 8) as well as on the demand for all types of transport and communications (see Chapters 9–11). At the same time changes in the overall structure of Europe's population are likely to have an important impact on the nature of household formation (see Chapter 4) and on the extent of labour force participation (see Chapter 5). With these considerations in mind this chapter reviews trends in the size and age composition of the resident populations of the European countries with particular reference to the three main components of population change: fertility, mortality and migration.

Trends

The most important demographic trend in recent years in most European countries is a decline in fertility. In the United Kingdom, for example, total period fertility rates fell from 2.79 in 1966 to 1.78 in 1986 (see Figure 3.1). In the Netherlands the decline was even steeper. In West Germany total period fertility declined from 2.53 in 1966 to a record low of 1.36 in 1986. This decline is not limited to north European countries. For example, the total period fertility rates in Italy fell from 2.62 in 1966 to 1.42 in 1985 and from 2.91 to 1.65 in Spain over the same period.

The decline in fertility is particularly marked in the 15 to 29 age-group. There is some evidence of increased levels of fertility in the 30+ age-groups especially in north European countries such as UK, the Nether-

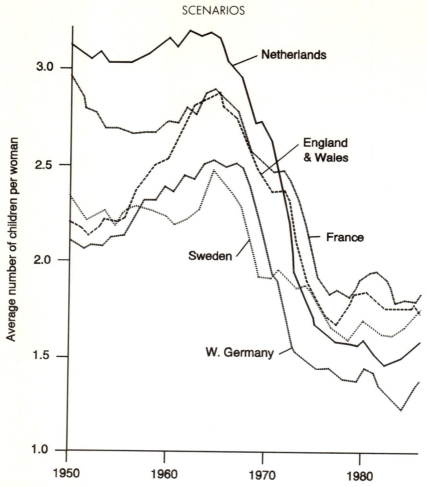

Source: Adapted from Bourgeois-Pichat, 1981

Figure 3.1 Birth rates in selected European countries. Fertility has declined dramatically in most European countries over the last 25 years. As a result very few countries have high enough levels of fertility to maintain their existing populations.

lands and Sweden. However, so far this looks as if it would only offset to a minor degree the overall decline in fertility over time.

Without taking account of international migration, a total period fertility rate of about 2.00 is needed to sustain a population at a given level over time. Given the above figures, most European countries are likely to experience a fall in population until 2020 (see Figure 3.2). The decline is likely to be most pronounced in countries such as Germany where the fertility rate

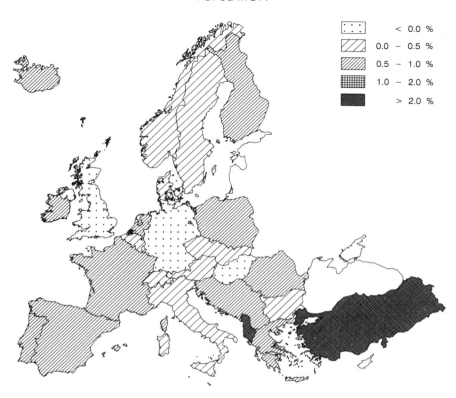

Figure 3.2 Population change in Europe, 1980–86. There are considerable differences within Europe in the current rates of population change which reflect their recent demographic histories. Consequently, while Germany, Hungary and the United Kingdom have declining populations, the population of Turkey is still increasing by more than 2 per cent per annum.

has already reached a very low level. A (pre-unification) population projection for West Germany suggested that the total population might fall more than 18 per cent between 1990 and 2020 from 61 to 49.9 million.

Another factor that must be taken into account is the impact of AIDS on mortality levels during the period under review. So far, AIDS in Europe is most marked among homosexuals and drug users, but its incidence among heterosexuals is increasing. By 2020 the impact of AIDS on overall mortality levels could be considerable unless some cure for the disease is generally available. This would lead to further population decline in most European countries.

The most fundamental impact of declining birth rates—in conjunction with increasing life expectancy—is the progressive ageing of the population. It is estimated that in western Europe the proportion of persons over 65 years will increase from 13 per cent in 1985 to more than 20 per cent in 2020, with increasing tendency beyond that year. Again, countries like Sweden and Germany are likely to be the leaders, surpassed only in the world by Japan (see Figure 3.3). The dramatic transition going on is illustrated by the changing shape of the age pyramid in Germany over time (see Figure 3.4).

The ageing of the population throughout Europe is likely to have important impacts on social and economic life in the twenty-first century. A considerable increase will be required in health and social services in all countries to meet the demands of old people. There will also be relatively fewer children in schools and there will be a surplus of jobs for school leavers. As the proportion of elderly in the population rises, there will also be increased demands on public transport and a relative decline in private car usage, especially in urban areas, unless new types of cars suited for elderly people are developed (see Chapter 10).

If only present trends are taken into account, it is unlikely that the population decline in most countries would be offset to any significant extent by international migration. The number of guest workers in most European countries has declined over recent years and restrictions on Commonwealth immigration in the UK have reduced the flow of immigrants to

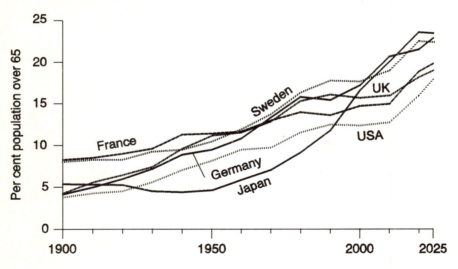

Source: Japan Institute for Social and Economic Affairs, 1989

Figure 3.3 Per cent population 65 years and older, 1900–2025. The ageing of the population is not a recent phenomenon. It has been evident in many European countries since the turn of the century. However, as the diagram shows, a considerable upsurge in the proportion of old people in the population after 2010 is expected as the effect of the boost to fertility provided by the post-war baby boomers diminishes.

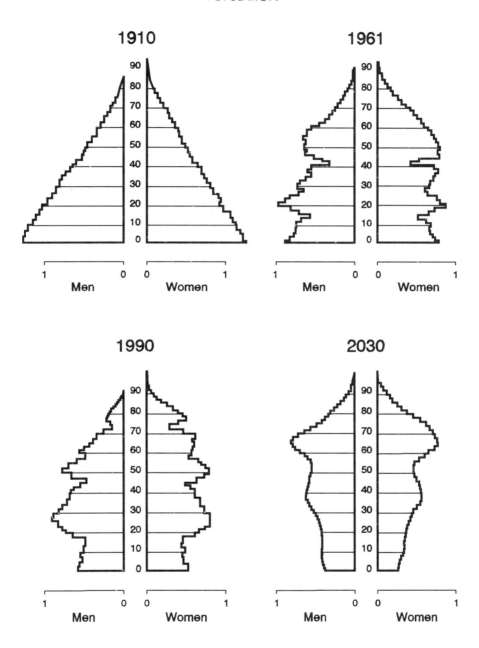

Figure 3.4 Age structure in West Germany, 1910–2030 (per cent). The changing shape of the population pyramid for Germany highlights the long-term impact of declining fertility on population structure. In this case it is predicted that the proportion of old people in the population will increase from 15.6 per cent in 1984 to 28.3 per cent by 2030.

that country. Nevertheless, work-related immigrations into the major receiving countries are still substantial.

Moreover, the impact of recent developments in eastern Europe could well spark a new wave of international population movements. Germany would be the most obvious beneficiary from such developments, but all west European countries may be affected.

In addition there remains the more fundamental question as to whether the affluent European countries will be able to shield themselves against immigration from their less affluent southern neighbours where the demographic transition has not yet started, i.e. population growth is still high. Table 3.1 highlights the enormous differences in population change between some EC countries and their neighbours around the Mediterranean. With few exceptions these countries will double or triple in population without a corresponding growth in their economy. It is hard to imagine

Table 3.1 North-South divide: population forecasts for EC and Mediterranean countries, 1985–2020

	Reality 1985	Forecast 2000	Forecast 2020	Change 1985–2020
	Million	Million	Million	Per cent
France	55	59	61	+10
West Germany	61	59	55	−11
Italy	57	57	54	−6
Portugal	10	10	10	+1
Spain	39	40	41	+6
United Kingdom	57	57	57	+1
Subtotal	279	283	278	−0
Algeria	22	34	52	+141
Egypt	48	66	92	+92
Lebanon	3	3	4	+38
Libya	4	6	12	+221
Morocco	22	32	44	+100
Syria	10	18	33	+212
Tunisia	7	10	13	+79
Turkey	50	68	86	+70
Subtotal	166	237	335	+102

Source: World Bank, 1990

Note: In contrast to the modest increases or declines that are expected in most European countries, if present trends continue until 2020, a substantial increase in population is anticipated in most African and Asian countries that border on the Mediterranean. As a result the balance of population between these two groups of countries is reversed during the period.

that the countries of the European Community will be able to completely deny their people access in a time of open borders and international cooperation.

Scenarios

The trends review highlights the decline in fertility throughout Europe and its impact on population growth. It suggests that there will be a general ageing of population which will impose severe strains on social welfare systems unless there is a massive influx of immigrants from eastern Europe and the Mediterranean.

Which of these trends will be dominant and which will have the greatest impact on the future population of Europe? Which of the resulting scenarios will be the most or the least desirable? The seed scenarios contained in Scenario Box 3.1 represent three contrasting positions.

The *growth scenario* is primarily concerned with maintaining and improving standards of living in an ageing society. This requires a heavy investment in measures to increase the productivity of the labour force and a relaxation of immigration restrictions for workers from developing countries. However, immigration is seen as essentially a temporary measure. As in the 60s and 70s these immigrants are essentially guest workers who are not eligible for permanent European citizenship. In this way the growth scenario enjoys the economic benefits associated with the net addition to the labour force without incurring most of the social costs of the migrants and their families.

The *equity scenario* revolves around the need to stabilize the ageing of the population of the European countries to avoid a crisis in social welfare delivery. In this case the most effective solution is to encourage immigration on a large scale from developing countries. Unlike in the growth scenario, these immigrants are eligible for permanent European citizenship if they wish to set up residence there. In this way the 'generation contract' where the young and active pay for the pensions of the elderly is maintained and population decline gives way to population growth on a scale unknown over the last fifty years. This policy would have the added advantage of contributing to reducing the disparities between the developed and developing countries.

The *environmental scenario* is built around a zero-growth population policy to reduce pressure on environmental resources, land prices and traffic congestion. In this scenario, maintaining current standards of living is likely to present formidable problems unless there are fundamental changes in social organization. Consequently this scenario envisages that, contrary to current trends (see Chapter 4), an increasing proportion of the

A   *Growth scenario*

By 2020, one quarter of the population in central Europe is over sixty. Catering for the elderly is the fastest growing market in the economy. Growing life expectancy and improved health have enabled elderly people to participate in sports, leisure and travel longer (see *Life-styles*). However, only continued growth in productivity (see *Economy*) makes it possible for the shrinking number of economically active to support the growing number of pensioners without a loss in standard of living—there is no alternative to growth especially for the richest nations. The real danger for a growth-oriented economy is the shortage of labour due to population decline. As measures to promote fertility have shown only little success, the government has relaxed immigration restrictions for workers from non-EC countries, but without granting them permanent European citizenship.

B   *Equity scenario*

In 2010, the European social security system underwent a deep crisis. Modelled after the West German and Swedish pension systems, it was built on the 'generation contract' where the economically active pay for the pensions of the elderly. With the growing number of old people and the decline of workers paying fees, the fund nearly went bankrupt. At the same time, the demand for health and social services grew while the taxable working population decreased. The way out was to encourage immigration from non-EC countries with growing populations. Like America two hundred years ago, Europe 2020 has become an immigration continent. The United Nations had long appealed to the rich nations to open their borders and share their affluence with the poor; this helped to stabilize the existence of the ageing European societies (see *Urban and rural form*).

C   *Environment scenario*

Under the zero-growth policy of European government (see *Economy*), the progressively ageing population in central European countries has become a serious social problem. Even to maintain the standard of living for the growing number of elderly people has become a heavy burden for the declining number of people of working age. Consequently, a growing number of old people has to be supported by their children. However, the decline in population has also reduced the pressure on environmental resources, on land prices in cities and on traffic congestion (see *Passenger transport*). Because the government supports young families (see *Life-styles*) many former working women have returned to the home, so conditions have become more favourable to raising children and fertility rates have started to go up in Germany, Holland and Sweden, while they continue to fall in most other countries.

elderly will have to be supported by their children. Similarly, if there is strong government support for young families, it is anticipated that many working women will return to the home.

The views of the experts

Are these the three basic options for population in Europe? On the whole the experts felt that the three scenarios covered the range of possible future developments. Nevertheless, they made a number of suggestions regarding possible modifications and requirements, particularly with respect to the key issues of the consequences of an ageing population, the scale of immigration from outside Europe and the extent of fertility decline.

The consequences of an ageing population

Given the pressures on the European social security system caused by the ageing of the population, several respondents from the NECTAR group saw a rise in the age from which state pensions are paid well before 2020. There was also general agreement that individual provision for pensions is likely to increase to offset the relative reduction in level of state pensions. Developments of this kind will tend to increase income disparities within the population. In the words of an English respondent, 'the effect of the changing age balance on the welfare of the older generations is considerable. There will be a divergence between the better-off and the less well-off who are dependent on social provision financed by the generation contract. The former, with the benefits of better health and longer life expectancy, can maintain their at-work standards of living, thereby blurring the distinction between the active and the retired by choosing to work longer with the possibility of reduced hours, working from home and the pursuit of second careers'. Similarly, a Dutch respondent predicted that by 2020 over half of the population aged over 60 will have to be supported to a considerable extent by their children while many other old people are able to spend a great deal of their time on exclusive leisure activities such as travelling to exotic places.

Some of the more optimistic among the respondents felt that the pressures on state social security systems created as a result of the ageing of the population will be relieved as a result of technological developments and increased labour productivity. They recognized, however, that this will require much more investment in technological innovation to enable machines to take over from humans in many areas of production.

The scale of immigration from outside Europe

There was a substantial body of agreement among the respondents with the view contained in the equity scenario that Europe will become an immigration continent by 2020. This is likely to have a profound impact on European society. As one Dutch commentator put it, 'there may be a massive population explosion in which non-Europeans dominate because of their higher fertility levels. At the same time, however, as noted above, birth rates in European countries may rise again so that the current double ageing process caused by ageing population and declining fertility is transformed into a double "younging" process created by an increasingly young population with high levels of fertility'.

There was a great deal of concern regarding the likely social consequences of immigration on this scale. As a result of more liberal immigration policies and the relaxation of immigration restrictions, a French commentator felt that 'there will be a difficult problem of co-habitation between the European population and the migrants especially in urban areas. To reduce the tensions that this produces governments will try to limit immigration for political reasons. This could be a source of very serious conflict between the European and north African countries'.

Many respondents doubted whether immigration restrictions could be enforced even if they were imposed by European countries. The likelihood that illegal immigration from eastern Europe and Africa might considerably exceed the number of jobs available should also not be ruled out. Even if this is not the case, immigration on this scale will have a marked effect upon European countries such as Portugal, Spain, Greece and Turkey who have traditionally acted as suppliers of labour to the richer countries of northern and western Europe. To tackle these problems, several respondents felt that it will be necessary for Europe to break free from its Eurocentric understanding of the world with all this implies in terms of explicit and implicit racism. It will be essential, therefore, to provide educational programmes and raise levels of awareness of different cultures to overcome racist and xenophobic attitudes.

The extent of fertility decline

Many respondents questioned the likelihood of further fertility decline. It was argued that in France, for example, there is some evidence to suggest that current figures overestimate the extent of fertility decline because of the degree to which many women are delaying having children. In some countries it was claimed that fertility rates will stabilize at a rather higher level in the future with longer gaps between generations due to the increase in the number of births to women aged over 30.

The degree to which fertility decline is likely to be halted or reversed is closely linked to the extent of female participation in the labour force in the eyes of many respondents. These see women as a significantly under-utilized resource in the labour force at the present time and predict that further incentives will be developed to encourage them to stay in work through, for example, flexible working hours, while those in part-time work will be encouraged to take up permanent appointments. In the words of an Austrian respondent, 'beyond all doubt conditions for maternity and child education have to be improved to maintain the actual level of birth rates or increase them slightly. But one cannot push women back to powerless domesticity, because as a result of the free access of young women to nearly all kinds of education since the 1960s, women like men gain their ego and economic independence through their participation on the labour market'.

Possible measures to stimulate fertility and make European society more children-friendly on the assumption that work for women and children can be made compatible include tax incentives, day care and crèche facilities for young children and sabbaticals for parents. It was also noted that in France, for example, school schedules are being rearranged to facilitate women's access to work.

Which scenario?

Despite the concern expressed above about the extent to which governments will be able to control immigration from outside Europe, the majority of NECTAR respondents (37) supported the view that the growth scenario will prevail throughout Europe by 2020. However, opinion was evenly divided between the three scenarios, environment (19), growth (18) and equity (14), when respondents were asked which they preferred for their own countries. This largely reflects the interpretations summarized above.

Speculation

Many of the elements of the growth scenario can already be seen in core countries such as Germany and the Netherlands. The extent to which this scenario might apply to other European countries by 2020 is debatable. Nevertheless, the underlying trend in this scenario, the ageing of the population, is also apparent in both the Scandinavian and the Mediterranean countries. Consequently, it can be argued that it is only a matter of time before countries such as Italy, Spain and Portugal move from a surplus of labour which is exported to the core European countries to a shortage of labour which has to be made up by immigration from outside Europe.

Recent work at the International Institute for Applied Systems Analysis (Lutz, 1991) suggests that the scale of immigration required to reverse the overall trend towards an ageing population would have to be massive (Table 3.2). In effect, nothing short of a social revolution is likely to reverse this trend. Such a revolution could bring a new vitality and dynamism to Europe but it would have very high social costs. It is also likely that its impacts would differ markedly from one country to another and from core to periphery.

The environment scenario requires the biggest changes in attitudes among Europe's population and the highest level of government intervention among the three population scenarios. Because of this it is the least homogeneous of the three scenarios and considerable variations can be expected not only between countries but also between different regions and different population groups within each country.

Under these circumstances policy-makers will have to choose between two courses:

either   to continue to permit non-permanent immigration of workers from non-EC countries without giving them full citizen status as suggested in the growth scenario. Such a policy would relieve the shortage of labour due to population decline in some European countries but would permanently establish a society based on discrimination;

or   to accept that Europe will become an immigration continent and to permit controlled permanent immigration from non-EC countries with growing populations as suggested in the equity scenario. Combined with a policy to support families and working parents as in the environment scenario, this would help to offset the ageing of the population and produce a more balanced age structure and labour market and would help to reduce the gap between the rich and poor countries.

The main choice relates to immigration. However, in practice the issues underlying this choice will be accentuated or reduced by the effects of any changes in fertility or mortality. A further decline in fertility and/or a further increase in mortality would increase the pressure on Europe's decision-makers to encourage or deter immigration.

Table 3.2 Immigration scenarios for West Germany

a. Total population (millions)

Annual number of immigrants	Size of population		Long-run stationary population
	1983	2033	
0	61.3	43.9	0.0
50,000	61.3	48.5	14.4
100,000	61.3	53.1	28.7
200,000	61.3	62.4	57.4
500,000	61.3	90.0	143.6

b. Foreign population (millions)

Annual number of immigrants	First-generation foreign residents		First and second-generation foreign residents	
	1993	2033	1993	2033
0	6.0	3.3	8.8	8.4
50,000	6.6	6.8	9.8	15.4
100,000	7.1	9.6	10.7	21.2
200,000	8.7	14.1	12.5	30.2
500,000	11.4	22.0	17.5	46.2

Source: Steinmann, 1991

Note: At least 200,000 immigrants every year for the foreseeable future would be required to maintain West Germany's population at its 1983 level according to forecasts made at the International Institute for Applied Systems Analysis. The social consequences of sustained immigration on this scale are considerable. By 1993 it is estimated that first-generation foreign residents would already account for 8 per cent of the total population. By 2033 this proportion would have increased to 14 per cent. If both first and second-generation foreign residents are taken into account, the proportions rise to 12 per cent and 30 per cent respectively.

Further reading

The twenty-five papers contained in the IIASA book on *Future Demographic Trends in Europe and North America* (Lutz, 1991) provide an authoritative overview of the whole field of population. The collection is divided into separate sections entitled 'The future of longevity', 'The future of reproduction', 'The future of migratory flows and regional trends' and 'Combining the three components in population projection'. These cover most of the issues raised in this chapter. For accounts of recent demographic developments in various European states the reader is referred to the publications of the Council of Europe on this topic (see, for example, Council of Europe, 1990) and to the volume edited by Findlay and White (1986).

An overview of mortality trends in industrialized countries can be found in Bourgeois-Pichat's (1984) study. The impacts of increasing longevity are explored in useful papers by Munton (1986) and Keyfitz (1989). A paper by Heilig in the IIASA book evaluates the impacts of AIDS on future mortality.

Coale and Watkins' (1986) study is the best introduction to the mechanisms of fertility decline in Europe. The main features of these trends are summarized in Werner (1988). The policy implications of fertility decline are explored in a special supplement of the *Population and Development Review* (Davis et al., 1986), and more recently in Wattenberg's (1989) book on *The Birth Dearth*. For those interested in the prospect for raising fertility in advanced economies, Hoem's (1990) review of social policy and recent fertility changes in Sweden contains some useful information.

An overview of current trends in migration in Europe can be found in *Contemporary Studies in Migration* (White and van der Knaap, 1985). The issues associated with the migration of the elderly are dealt with in some detail in the volume edited by Rogers and Serow (1988). Some of the social problems arising out of the migration of guest workers to the core European countries are discussed by King (1990). The future development of labour markets in Europe up to 2000 is discussed in a Prognos report (1989) and some of the effects of the Single European Market on labour migration are explored in a paper by Martin et al. (1990).

CHAPTER 4

LIFE-STYLES

Definition

The range of life-styles of Europe's population is the product of changing social values, the dynamics of population growth and decline (see Chapter 3) and the varying demands of local and regional economies (see Chapter 5). The main features of these life-styles are particularly apparent in the size and composition of households and in the daily activity patterns of the individuals who make up these households. These in turn are reflected in the level of demand for transport and communication facilities of all kinds (see Chapters 9–11), the nature of urban and regional development (see Chapters 7–8) and the pressures that are placed on environmental resources (see Chapter 6). With these considerations in mind this chapter reviews trends in household size and composition, labour force participation, work and non-work time regimes and activity patterns with particular reference to the demographic, social and economic factors underlying them.

Trends

Over the last thirty years the size of the average household has fallen substantially in all European countries (see Figure 4.1). In Great Britain, for example, the average household size fell from 3.1 persons in 1961 to 2.6 in 1986. Over the same period the average size of households in Germany fell from 2.8 to 2.3. These trends are not confined to the north European countries. In Italy, for example, the average household size declined from 3.6 to 2.8 over the same period.

These trends reflect a number of demographic, social and economic factors which have influenced life-styles in European countries (Hall, 1988). The most important demographic factors influencing household size are

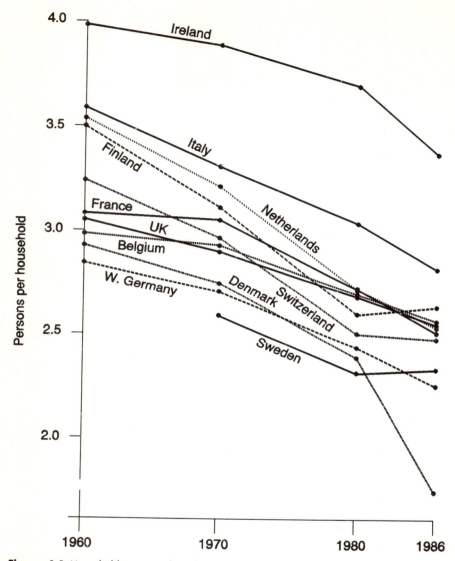

Figure 4.1 Household size in selected European countries, 1960–86. There has been a dramatic fall in average household size in most European countries over the last thirty years. This reflects not only demographic factors associated with the ageing of the population but also social and economic factors related to the breakup of the extended family system and the growing economic independence of many young people.

the decline in overall fertility over this period which has led to much smaller families and an increase in the proportion of old people in the population (see Chapter 3).

The most important social factors are the decline in three-generation families and the reduction in the proportion of married couples in the population. During the 1970s, divorce rates doubled in Belgium and France, trebled in the Netherlands and quintupled in England and Wales. By 1986 births outside marriage accounted for nearly half the total births in Denmark and Sweden.

The most important economic factors are increasing overall affluence, the growing economic independence of women and young people, and the decline of traditional peasant agriculture. With respect to the latter, in Italy, for example, the agricultural labour force declined from 17 per cent in 1970 to 11 per cent in 1985.

The overall impact of these demographic, social and economic trends is a reduction in the number of large households and an increase in the number of single-person households. In Great Britain the proportion of households with five and more members fell from 14 to 11 per cent between 1971 and 1981, while the proportion of single-person households rose from 18 to 22 per cent. The corresponding figures for West Germany show a reduction from 13 to 8 per cent in households with five or more members and an increase in one-person households from 25 to 31 per cent. In Italy households with five or more members decreased from 22 to 15 per cent over this period, while single-person households increased from 13 to 28 per cent.

Further declines in average household size can be anticipated between 1990 and 2020 as the effects of demographic, social and economic factors work through the population. The impact of such trends is likely to be particularly pronounced in wealthy countries with very low fertility and ageing populations such as Germany, but most apparent in absolute terms in southern European countries such as Italy where fertility levels are falling and traditional peasant societies are being replaced by urban cultures.

With households becoming smaller, more women tend to work. In most countries labour force participation increased between 1976 and 1985 due largely to the increase in number of working women. In the UK, for example, overall activity rates rose by 5 per cent over that period to 60 per cent. Female activity rates increased by 12 per cent to 48 per cent while male activity rates remained at 73 per cent. The increase in female labour force participation was particularly marked in southern countries such as Italy where female activity rates rose by 20 per cent to 34 per cent between 1976 and 1985. During the same period male activity rates increased only a little to 67 per cent. Despite these similarities there are still large differences in labour force participation between the European countries (see Table 4.1), some of which are due to cultural attitudes and traditions.

Another factor works in the opposite direction, i.e. decreases labour force participation. As societies become more affluent, social security and pension schemes make it unnecessary for old people to work for their live-

Table 4.1 Labour force participation rates, 1986

| | Labour force participation (per cent) | | | |
	Male	Female	65+	Total
Austria	70.6	40.6	1.5	54.6
Belgium	62.5	36.2	1.7	48.9
Denmark	73.6	60.1	7.7	66.7
France	67.3	46.2	2.5	56.2
Germany, F.R.	69.9	41.0	3.2	54.6
Greece	67.8	34.0	10.3	50.1
Ireland	71.2	32.9	10.6	52.0
Italy	66.8	33.5	4.9	49.5
Netherlands	66.6	34.5	2.4	50.4
Portugal	72.2	45.0	12.3	57.8
Spain	66.6	27.1	3.6	46.0
United Kingdom	72.5	48.4	4.8	60.0

Source: Eurostat, 1988
Note: Although female labour force participation rates are generally rising throughout Europe, there are still large differences between countries which reflect cultural rather than economic factors. For example, female labour force participation in The Netherlands is only two-thirds of that of the United Kingdom.

lihood. This creates financial problems in an ageing society because a smaller number of economically active people have to support a growing number of pensioners, but that has nowhere led to a reversal of the decline in labour force participation of elderly people. Table 4.1 testifies that the labour force participation of people over sixty-five is lowest in the more affluent countries of Europe.

The proliferation of small households and new patterns of labour force participation have created new life-styles, which not only reflect the particular attitudes of a certain generation and period, but also have a profound impact on settlement systems and urban form. Up to 80 per cent of all households in inner cities are singles: either workers, unemployed people, students or pensioners, who live in the inner city because they depend on cheap, run-down housing and have to rely on public transport, or yuppies ('young urban professionals') who prefer luxury flats and are indispensable for new up-market shops and restaurants. Dinks ('double income no kids') are childless couples who both work and have the spending power for second homes or expensive holidays. A third important new life-style is that of the 'silver' generation, the early retired, relatively affluent 'senior citizens' described in Chapter 3, who are still healthy enough to enjoy spending their money on housing, recreation, entertainment and travel; they represent a sizeable and growing market.

The darker side of the demise of the large family is the fragmentation of the activity spheres of the individual. The liberation from the traditional bonds of family, neighbourhood or community may facilitate a richer set of transient attachments for the young and active at the same time as cultural diversity is increasing as a result of greater international mobility, but it can also mean isolation and loneliness for the old and sick, polarization of interests and discrimination against ethnic, cultural or social minorities. The ideal of the pluralist, multicultural, tolerant and cooperative society has yet to demonstrate its viability.

Another aspect of the ageing society, but more importantly of new technologies in manufacturing and services (see Chapter 5), is a marked increase in free time. If automation increases productivity faster than output grows, labour becomes redundant. This can lead to unemployment, but can also be turned into reductions of working hours per day, per week, per year or per lifetime. This has happened in all industrial countries since the abolition of Sunday work, and although there are still large differences between the work time regimes in individual European countries, it is safe to say that by the year 2020 both the eight-hour workday and the five-day work week will be a thing of the past (see Figure 4.2). Also there will be a great diversity of work hour arrangements ranging from flexitime to teleworking from the home via computer link (see Text Box 4.1).

The price to be paid for all this will be a certain fragmentation and unpredictability of work schedules subject to the requirements of continuous production processes or shop opening hours, but also a new flexibility and scope for self-determination, a trend accentuated already today by the increase in the number of workers involved in job sharing, sabbaticals and early retirement programmes. A consequence of this will be that the conventional distinction between work-related activities and non-work activities will begin to break down and there will be an increase in informal work arrangements and a corresponding growth of the black economy. More people will also find it necessary to have two or three jobs to sustain their particular life-styles.

One of the effects will be an enormous increase in the amount of time devoted to leisure activities. To meet these demands, it is likely that there will be a massive expansion of the leisure industry (see Figure 4.3). By 2020 leisure activities may account for as much as 40 per cent of all land transport (in terms of kilometres travelled) and 60 per cent of air transport. The growing diversity of life-styles will be reflected in the emergence of new types of specialist tourist markets catering for young adults and couples without children, young retired people and conference delegates. The growing demand for tourism will be particularly evident in the population of the southern European countries where current participation rates are relatively low. A further boost to international tourism can be expected after 1992 with the removal of many of the existing institutional

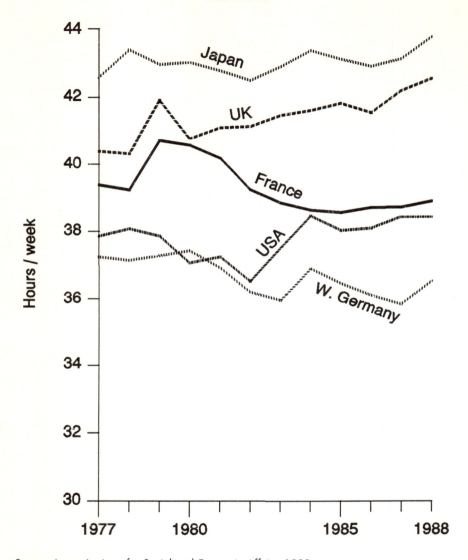

Source: Japan Institute for Social and Economic Affairs, 1989

Figure 4.2 Hours actually worked in manufacturing industry (production workers) 1977–88. There are considerable differences between countries in the trends with respect to the hours worked per week in the manufacturing industry. While the numbers of hours worked by British workers appears to be increasing slightly over time, those of French and German workers have substantially decreased since 1977.

F INTERNATIONAL

F International is a systems house which has people as its main asset, is known as the company without offices (in fact it does have some) and is an integral part of the world of information technology. I founded the company back in 1962 and it has prospered to become the world leader in distributed office work. I define office work as revolving around the management, administration and servicing of organizations and their customers and users. My approach to the distributed office concentrates on practical examples of office work carried out from home via the telephone. The term used is *telework* and my claim to be a *teleworker* is because I have worked from home for 25 years on a variety of office type tasks from technical supervision to strategic management.

My own F International came to the overheads problem from a new angle: how to access a skilled workforce in short supply. It trades primarily in Northern Europe and is a computer systems and software company targeted at four market sectors: financial services, retail and distribution, science and engineering and community (mainly health). The company operates as a wholly teleworking establishment and its Mission, 'to develop, through modern telecommunications, the utilized intellectual energy of people unable to work in a conventional environment' has been implemented throughout the company since 1962.

The company has teleworked through the various stages of the organization's development, from entrepreneurial mode through semi-autonomous regions on to its current maturity as a managed organization. Similarly, it has teleworked through various phases of the technology; from the simple telephone through the on-line stage to the current era where the technology and its decreasing cost allows homeworkers to be increasingly equipped with electronic mail and similar facilities.

F International therefore offers quite a degree of generality! And one which can be measured in commercial terms. It is ranked 12 in the UK league of software houses, the industry being very fragmented, but represents a £6 billion annual turnover and employs over a quarter of a million people. F International has 1,000 people of whom 250 are employed and 800 are co-working associates under regular sub-contract. Nearly all work from home.

Contrary to expectation, the technology is the least pressing concern. It is the human interfaces which are critical. Teleworkers provide business flexibility—a key to economic survival today. People are available as and when needed and generally paid in terms of productive output rather than presence. Those covering essential functions, either for F International or its customers, can work 'unsocial hours' if it suits them. Others have a portfolio of activities perhaps doing seasonal work, rearing children, studying or in a variety of ways finding a life-style which suits their family unit. A few are disabled or looking after a disabled member of the family, usually an elderly parent. Staff turnover is low for the industry because the workplace is immaterial and people are not lost to the company when a partner has to move location, nor their time lost due to immobilizing illnesses, severe weather conditions or national emergencies like the 3-day week.

Text Box 4.1 The case of F International (Shirley, 1987). This is a good example of the new forms of employment that are emerging as a result of recent developments in communications technology, changes in life-styles and new types of organization.

barriers to movement. This may also result in an increase in the number of young retired people from northern Europe moving to southern European countries to take advantage of the favourable climate. It is also likely that many people will use their extra free time to acquire new skills and broaden their ranges of experience. This will be reflected in a rise in the number of mature students attending places of higher education.

Scenarios

The most important feature of current trends is the fall in the average size of household throughout Europe. This reflects social and economic as well as demographic factors. As a result there has been an increase in female labour force participation and a proliferation of small households with a wide variety of life-styles. The decline in working hours will give rise to a massive expansion of leisure activities of all kinds.

What will be the impact of these trends on future life-styles in Europe? Three contrasting life-styles scenarios are portrayed in Scenario Box 4.1.

The *growth scenario* is dominated by the individual preferences of the better-off sections of the community. The role model of this scenario is the affluent and highly mobile young white-collar professional. The consumption of energy and natural resources associated with this life-style is considerable. Coupled with a massive overall increase in commercial leisure

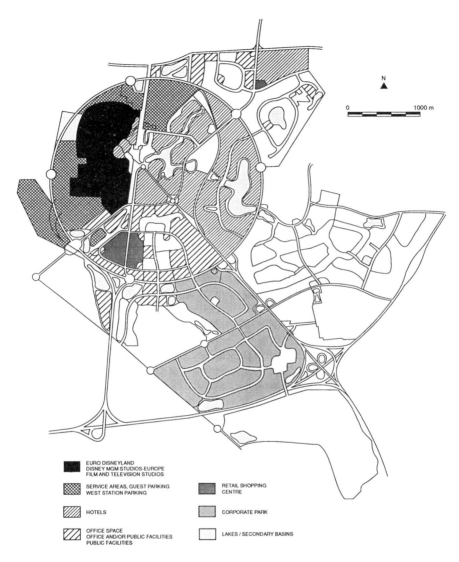

EURO DISNEYLAND
DISNEY MGM STUDIOS-EUROPE
FILM AND TELEVISION STUDIOS

SERVICE AREAS, GUEST PARKING
WEST STATION PARKING

RETAIL SHOPPING
CENTRE

HOTELS

CORPORATE PARK

OFFICE SPACE
OFFICE AND/OR PUBLIC FACILITIES
PUBLIC FACILITIES

LAKES / SECONDARY BASINS

Source: Euro Disney S.C.A., 1990

Figure 4.3 Euro Disneyland master plan. The Euro Disneyland resort which is under construction on a 1943 hectare site at Marne la Vallée 32 kilometres east of Paris illustrates the scale of investment that is involved in the development of a commercial leisure complex. When the project is completed in 2017 the site will contain not only two major theme parks (the first of which opens in April 1992) but also over 18,000 hotel rooms, 700,000 m² of office space, 95,000 m² of retail space, a large multipurpose corporate park and an international convention centre. The resort will be linked to Paris directly through an extension of the existing rapid transit (RER) system and will also have a link to the TGV high-speed rail network.

A   *Growth scenario*

Besides the elderly people, the singles and 'dinks' (double income no kids) are the engines of the growth economy of the early twenty-first century. Small households spend more on housing, travel, clothing and food, so the decline in household size helps to sustain the economy even at a time of population decline. The government does nothing to stop the fragmentation of the family, but encourages even single-parent families to remain in the labour force. The economically independent young white-collar worker is the model for the life-style of the decade, characterized by efficiency, high personal mobility (see *Passenger transport*), intensive use of telecommunication (see *Communications*) and vast consumption of energy and resources (see *Environment*). This scenario is dominated by individual and personal preferences. There is a massive growth of commercial leisure activities with amusement parks springing up everywhere. However, the success of the young and active rests on the declassification of the less able and efficient: the growth scenario has a dark side of poverty.

B   *Equity scenario*

In the early 2000s, a fundamental change of values takes place in Europe. Fifteen years of redistribution of wealth and a ruthless ideology of competition, extreme individualism and the fragmentation of social ties have widened the gap between the generations and between social and ethnic groups (see above). Now young people are alienated by the 'elbow mentality' of their elder generation and want a world of equal opportunity and social peace built on solidarity instead of competition. The family, having children, to be at home with friends and other traditional values have experienced an unexpected renaissance. Hand in hand with this there is a growth of collective as against individual life-styles; participation in community affairs, team building and consensus formation are emphasized; there is also a new interest in adult education.

C   *Environment scenario*

Life in harmony with nature requires mutual help and economic activities which are not market-regulated, and these can exist nowhere better than in a family. So the government promotes three-generation families by tax incentives and housing support for young people (see *Population*). Many of its measures are directed against the energy and resource consuming activities of the leisure world such as skiing or car driving (see *Environment*). This is of course resented by the car and leisure industries, but also by young people who have no experience of the ecological disasters of the 1990s and feel only deprived of their pleasure. A growing proportion of the extra leisure time gained by shorter work hours is spent on conservation activities, partly voluntary, partly in the form of compulsory community services.

facilities, the impacts of this scenario on environmentally sensitive areas such as the Alps and some parts of the Mediterranean could be disastrous. An important feature of this scenario is the extent to which the gap between the 'haves'—typified by the young white-collar workers, and the 'have-nots' exemplified by the ageing blue-collar workers, will increase, despite the increased productivity of the latter. The decline in working hours is also likely to be most marked for the latter who will have a great deal more leisure time without the necessary income to take advantage of the opportunities that are available.

The *equity scenario* is based on a rejection of the elbow mentality that is at the heart of the growth scenario. It envisages a return to a more caring society based on mutual support rather than competition. Important features of such a society would be a renaissance of family life and a growth of collective as against individual life-styles. Under these circumstances the decline in average household size would be reversed and the gap between the 'haves' and 'have-nots' reduced.

Some of the features of the equity scenario are also incorporated in the *environment scenario*. However, this also envisages various forms of community activities—both voluntary and compulsory—taking up an increasing proportion of the extra leisure time created by shorter working hours and more flexible work arrangements. Environmental conservation occupies a central place in this scenario which assumes that the experience of the ecological disasters in the present decade will result in a major shift in public opinion and lead to stringent measures against energy and resource-consuming activities such as skiing and motoring. In this way the pleasures of the young white-collar workers in particular will be curtailed in the interests of environmental conservation.

The views of the experts

Do these scenarios represent the basic life-style options for Europe in 2020? If so, what are their implications for future policy-making?

Although there was general agreement among the experts that the scenarios covered the range of potential developments, a number of modifications were suggested and several alternative scenarios put forward containing elements from one or more of these scenarios. These dealt primarily with the following issues: the impact of a greater diversity of life-styles, the nature of the renaissance of the family and possible changes in the dominant life-styles.

SCENARIOS

The impact of a greater diversity of life-styles

Although some respondents felt that the growing diversity of life-styles depicted in the growth scenario was fairly balanced and nested in the social attitudes of the majority of people, many of those questioned the extent to which the three scenarios were in fact mutually exclusive. In the words of an English commentator: 'all three component scenarios will coexist alongside each other embodied in different groups within the population'. This is an important trend in itself reflecting a transition from relatively homogenized social mores to a much more diverse and fragmented societal expression. Similarly, a Swedish respondent commented that 'the scenarios represent significant options of life-styles. One further option could be a mixture of the three where young white-collar workers, young people alienated from the elbow mentality and people in harmony will stress their own life-styles'.

The extent to which middle-class values and ideologies tend to dominate the scenarios was noted by several respondents, who felt that more attention should be given to elaborating scenarios relating to the dark side of poverty. Some respondents also felt that the scenarios reflect the European aspects of changing life-styles and that it will also be necessary to take account of locality-specific characteristics which will continue to remain important despite Europe's increasing multicultural character. It was also postulated that the great diversity of life-styles may be related to the economic transformation from Fordism (see Chapter 5) to post-Fordism in that the partial demise of the system of mass production and its associated mass consumption has put in train parallel changes in ways of life possibly signifying a shift towards a distinctly post-modern cultural ethos.

The nature of the renaissance of the family

There was a great deal of support for the sentiments expressed by a French commentator regarding the forthcoming renaissance of the family in Europe. 'Yuppies and dinks are not permanent status symbols but trade marks of the 1970s and 1980s. Family values are coming back even in a liberal and selfish society because children who have suffered from broken families will react against this tendency'. Under these circumstances the values of communal solidarity which is strongly entrenched, for example, in Jewish people, are likely to be enhanced and there will be a return to fundamentalism in society.

Nevertheless, many commentators pointed out that the renaissance of the family does not necessarily imply a return to Victorian values but is more likely to reflect the emergence of new styles of living. In the words of a Swiss respondent 'new forms of living will be developed and gain importance. Big households need not necessarily consist of one family only. Old people may live together or some one-parent families. Household sizes will vary a lot more than in the 1990s. If governments encourage big households, small households may come to live together sharing facilities such as cars, TV rooms and laundry equipment'. A greater flexibility in household arrangements is likely to be an important feature of these new lifestyles which will combine a considerable amount of freedom for individual members while promoting at the same time some measure of communal solidarity through, for example, the elderly looking after the children.

Another factor that may help to promote greater communal solidarity is the emergence of new forms of work. The extent to which there will be a decline in working hours was questioned by a Swedish respondent on the following grounds. 'Actual working hours in manufacturing industries are declining, but working hours in the private sector are increasing. The net result might be stable or increasing working hours. The voluntary exchange of services among friends will also tend to increase actual working hours'.

Changes in the dominant life-styles

Many respondents saw a major change in direction of the dominant forms of life-style after 1990. According to one English commentator 'it is happening already, due to a recognition that the unfettered individualism of the 1980s in the UK at least is not only socially unsustainable but is also regressive in economic development terms'. In the opinion of a Dutch respondent 'children who will be born in the last decade of the twentieth century won't like the society created by their parents very much. So they will arrive at the equity scenario automatically'.

A number of respondents saw ecological factors as playing a major part in producing a change of direction of life-styles. In the eyes of a French commentator 'the present growth scenario is not sustainable in the medium to long-term perspective because of environmental and energy constraints. So there will be a dramatic change in the near future'. Similarly, an Austrian respondent claimed that 'the growth scenario is drawing to an early end because of the ecological barriers that hinder most consumption of energy and resources. Annual budgets will face a dramatic increase of costs for conservation activities. By 2020 the collective threat will reduce the elbow mentality, and responsible individualism will defend its place in a tolerant and cooperative society that has to be keen on experimenting at all levels: individual, household and community'.

Generally it was felt that a change in direction of this kind and the change in values that is involved can only come out by itself although governments may attempt to stimulate it by a variety of measures including tax incentives. In the opinion of one Austrian respondent, possible measures include 'the creation of basic facilities for young families and households (job sharing, kindergartens, time for leisure), financial recognition of housework and child education, equal status for men and women, improvement of public services (household services, public transport, education) with increasing public participation, open-minded regionalism and decentralization within a modern legal framework and acceptance of the ecological aspects relating to the quality of life'.

Which scenario?

The majority of respondents (38) thought that the growth scenario will prevail in 2020. However, this may have to be modified in a variety of ways to take account of the views expressed above particularly with respect to the degree to which elements of the other two scenarios can be incorporated in a growth scenario without changing its essential features. Nevertheless, when it came to their preferred scenarios, nearly half the respondents (25) favoured the equity scenario with its emphasis on the need for collective as well as individual life-styles and support for the renaissance of the family. Only a small minority (12) preferred the environment lifestyle scenario for their own countries.

Speculation

Many of the features of the growth scenario can already be seen in countries such as Germany and Britain. However, although the decline in household size is a common feature of most European countries, the extent to which it will give rise to increased female labour participation is likely to vary because of the differing cultural traditions between countries. The relative diversity of life-styles in the urban areas of some countries may still be markedly different from the relative homogeneity of rural life-styles in these countries even in 2020.

Because of these differences, many respondents foresaw that elements from each of the three scenarios may coexist alongside each other in some countries. Nevertheless, they showed a strong preference for the equity scenario even though its implementation would necessitate a major shift in social values. A shift from individual to collective goals is most likely to be achieved in circumstances where there is a strong consensus within society. Such a shift may present particular problems in the larger urban areas because of their great cultural diversity which is likely to be further increased by immigration (see Chapter 3). Probably the most important thing that governments can do to foster such a reorientation of values is to create new forms of organization which encourage collective decision-making at all levels.

There are many common elements in the equity and environment scenarios. Both stress the need for a major reorientation of values and a revival of family life. However, the distinctive feature of the environment scenario is the emphasis that is given to environmental conservation. Measures to curb activities such as skiing and recreational car use will not only require considerable self-discipline within society but will also have differential spatial impacts particularly on the Alpine countries and on some Mediterranean countries if restrictions are imposed on the use of beaches to limit sea pollution.

Given these circumstances, European decision-makers will have to choose between two options:

either   to accept the highly individualistic and resource-consuming society that is depicted in the growth scenario, together with its dark side of poverty and increasing disparities within society;

or   to promote some combination of the equity and environment scenarios and return to a collective and caring society accompanied by a renaissance of the family, albeit in forms very different from their Victorian forebears, to reverse the decline in household size.

The alternative to the individualistic scenario requires fundamental changes in current attitudes. Nevertheless, it is interesting to note that it was favoured by many of the younger respondents who are likely to be responsible for such decisions in a generation's time.

Further reading

The new life-styles that are emerging throughout the world are graphically described by Naisbitt and Aburdene (1990) in the chapters of *Megatrends 2000* on 'Global life-styles and cultural nationalism' and 'The triumph of the individual'. Some elements of the darker side of current changes in social structure are examined in the British context by Hamnett *et al.* (1989). Close and Collin's (1985) work gives a good introduction to family and economy in modern society.

Bell's (1973) *The Coming of Post-Industrial Society* anticipated many of the issues discussed in this chapter. Northcott *et al.* (1991) contains some interesting data on changes in expenditure, consumption and activity patterns in Britain.

Some of the main features of current trends in household change are summarized in a useful paper by Hall (1988). A note of caution regarding the interpretation of the current decline in household size is to be found in Wall's (1989) historical perspective on trends in European countries. The findings of a detailed analysis of cohabitation, marriage and separation in France between 1968 and 1985 are described in two papers by Leridon (1990a, 1990b).

A volume of essays edited by Pahl (1988) provides a good introduction to changing views on work. The consequences of increased female participation in the labour force are discussed by McDowell in her chapter of the Hamnett *et al.* (1989) book on the changing social structure of Britain. The work edited by Vianello and Siemienska (1990) contains the findings of an exhaustive comparative enquiry into gender inequality in Canada, Italy, Poland and Romania. The social implications of information technology are examined by Miles *et al.* (1988), and Kinsman (1987) contains some interesting case studies of telecommuting.

The volume edited by Williams and Shaw (1988) on *Tourism and Economic Development* gives a good overview of the experience of west European countries which complements the material contained in official reports such as those compiled by the OECD (1986a). A paper by Bieber and Potier (1992) sets out a number of scenarios for transport in relation to the development of tourism. For an insider's view of the tourist business the reader is advised to consult Lundberg's (1990) textbook.

CHAPTER 5

ECONOMY

Definition

All the other seed scenarios are likely to be affected to a considerable extent by the quantitative and qualitative changes that will take place in the European economy up to 2020. The scale of future economic growth will be manifested in the demand for all kinds of transport and communications (see Chapters 9–11). It will also have a profound impact on the trajectories of differential regional development within Europe (see Chapter 7) and on urban and rural form (see Chapter 8). There is also a strong connection between changing modes of employment and the development of particular life-styles (see Chapter 4). With these considerations in mind, recent economic trends in European countries are considered with particular reference to economic change, industrial organization and the forms of industrial reorganization that are likely to follow the completion of the Single European Market in 1993.

Trends

Throughout most of Europe there has been a decline in traditional manufacturing industries during the last decades. This has to a large extent been compensated for by the expansion in service activities. In the UK, for example, the proportion of the labour force employed in industry fell between 1976 and 1985 by 20 per cent to 32 per cent while the service sector increased its share by 11 per cent to 65 per cent of total employment (Figure 5.1).

Similar trends can be found in most European countries. In West Germany industrial employment over the same period fell by 9 per cent to 41 per cent of total employment, while the share of service employment

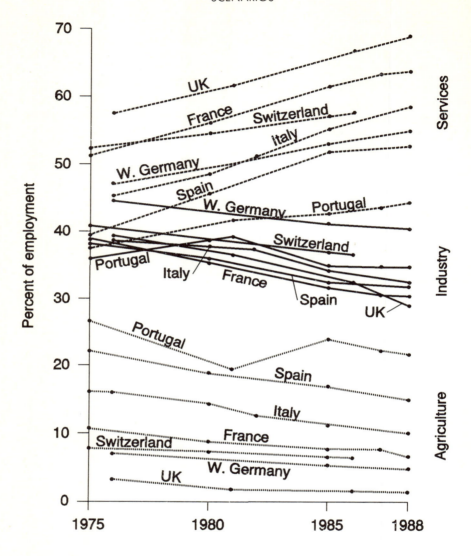

Figure 5.1 Agricultural, industrial and service employment in European countries in percentages, 1975–87. All the major countries exhibit a decline in agricultural and manufacturing employment since 1975 with a corresponding increase in the service sector. This is most pronounced in countries such as Britain and France but late starters such as Italy and Spain are rapidly catching up with the central European countries.

increased by 10 per cent to 53 per cent. In Italy the share of industrial employment fell by 8 per cent to 34 per cent while the share of the service sector rose by 28 per cent to 55 per cent. Throughout Europe agricultural employment declined in the same period. It fell by 8 per cent in the UK, by 17 per cent in West Germany and by 29 per cent in Italy. In Italy the proportion of the labour force employed in agriculture fell from 16 to 11 per cent between 1976 and 1985. In most countries the highest rates of growth over this period were recorded in information-related activities such as finance and insurance. In the UK employment in this sector rose by 32 per cent between 1976 and 1985, in West Germany by 22 per cent and in Italy by a staggering 71 per cent.

The trend scenario for the period up to 2020 assumes the continuation of structural economic change with corresponding decreases in employment in traditional manufacturing industries. In most north European countries agriculture is already a relatively minor source of employment which is declining steadily. In the southern European countries where agriculture is still a relatively large source of employment, the overall rate of decline is likely to be higher than in the north European countries.

However, the shift in sectoral composition of employment reflects only one aspect of the transition of the economic system to the 'post-industrial' society. Behind it there are more fundamental changes in the total organization of production and distribution which are described as the transition from 'Fordism' to 'post-Fordism'.

Fordism indicates the era of mass production dominated by economies of scale. With the gradual introduction of computerization in manufacturing (CAM), a new flexibility and responsiveness of the production process is achieved: economies of scale are replaced by economies of scope. This becomes possible by an increasing vertical integration of all steps of the production process from supply to delivery by computer control and telecommunications (see Chapter 11). At the same time earlier steps in the assembly chain are increasingly contracted out to outside suppliers who have to synchronize their operations and delivery with the main production schedule (just-in-time delivery). This has significant implications for spatial structure (see Chapter 8) and goods transport (see Chapter 9).

The transition to post-Fordism also changes the character of work. Ever more sophisticated machines take over more and more of the repetitious and monotonous tasks at the assembly line, so the part of the human becomes more supervisory. Where manual work is still required, the traditional division of labour is replaced by more comprehensive work packages to increase job satisfaction and responsibility. The proportion of jobs requiring higher skills is growing as is the number of staff employed in research and development and sales. Workers who are not able to adjust to the new skills are becoming redundant. For the remaining work-force, individual work hours are decreasing with rising productivity, but continuous production jobs require more shift work or work on Saturdays (see Chapter 4).

The growth in the service sector does not compensate for the industrial job losses. The surplus manufacturing workers do not have the skills required for the new high-level financial and consulting services nor can they compete with cheap temporary unskilled labour hired by retail and fast-food outlets. Especially in the wealthier countries, skilled personal services are becoming more and more unaffordable for the majority of the population, with the effect that the alleged 'service society' is gradually turning into a 'self-service society'.

Another characteristic of the post-industrial economy is a polarization of firm sizes. On the one hand there are very large corporations which continue to become even larger. In addition to the vertical integration referred to above, they employ a strategy of horizontal integration by acquiring smaller companies to diversify into fast-growing high-tech products or services. Large companies increasingly become multi or transnational in order to compete on the world market and to exploit labour cost differentials between individual countries (see Table 5.1). This may even lead to a separation of activities within one corporation over different countries. There is already a general tendency for high-level knowledge-based and

Table 5.1 Wage costs, productivity and unit costs in countries of the European Community, 1984

	Wage costs	Productivity	Unit costs
Belgium	106	107	99
Denmark	107	92	116
Germany, F.R.	124	107	116
Greece	38	58	66
France	107	110	97
Ireland	88	81	109
Italy	95	101	94
Luxembourg	97	110	88
Netherlands	117	132	89
Portugal	24	51	47
United Kingdom	87	92	95
EC (without Spain)	100	100	100

Source: Eurostat, 1988
Note: 1. EC = 100
2. There are considerable differences between the European countries with respect to wage costs, productivity and unit costs in 1984. Average wage costs in Portugal are less than one-fifth of those in Germany but productivity is also much lower. In other countries such as The Netherlands productivity is high relative to average wage costs, thereby ensuring below average unit costs.

skill-intensive tasks to concentrate in a few core cities and regions while low-skilled standardized production tasks are carried out in peripheral areas within individual countries and also within Europe as a whole (see Chapter 7). With the opening up of markets in eastern Europe, countries such as Poland and Hungary may increasingly become targets for these branch plant economies.

On the other hand there is a fast growing number of small companies. Empirical studies have shown that much of the employment growth in recent years has been due to small and medium-sized firms with innovative product or service ideas. In many countries special subsidy programmes have been set up to promote the establishment of new promising companies. Technology centres are set up as incubators for innovation-oriented enterprises, which later move out to new technology parks and enterprise zones to attract investors with reliefs from taxes and environmental regulations (see Text Box 5.1). Technopolis programmes imitating the Japanese example are other technology-oriented policies directed to small and medium-sized firms, although in most countries the bulk of high-tech promotion funds go to the big companies of the military-aerospace complex.

Despite the growing internationalization and integration of Europe there remain considerable disparities in economic strength between the European countries. The difference between relatively rich countries such as Germany and relatively poor countries such as Greece may increase rather than decrease in the early twenty-first century as a result of the opening up of the east European markets. The impacts of greater European integration after 1992 are likely to lead to a single European currency and a dramatic expansion in international exchange and collaborative ventures between industrial groups in the different European countries. In addition, the Community will grow. Already in 1993, the EC and the countries of the European Free Trade Association (EFTA) will join to become a free trade region of more than 375 million people (see Figure 5.2), and this will only be a prelude for closer links with other European countries ranging from loose association to full membership.

Throughout the whole period up to 2020 economic developments in Europe will be profoundly influenced by developments elsewhere in the world. The emergence of new superpowers in the less developed world, such as China, during this period will increase international trade competition and create new pressures on traditional markets for European products and services. The future of Europe as a trading partner is also likely to be closely linked to the extent that European interests are represented in the limited number of large multinational companies that will continue to dominate the world economy.

PROSPECTUS

Cambridge Science Park

* Established in 1970 by Trinity College, the college of Newton and many other distinguished scientists;
* Close links with the scientific excellence of Cambridge University;
* Low density development in a park-like setting;
* Some 80 compatible neighbours also involved in high technology;
* High quality, flexible buildings suited to office, laboratory or manufacturing use.

Use clause

* To preserve the intrinsic nature of the Cambridge Science Park for the mutual benefit of all occupiers, the use of buildings there is limited to the following:

  * scientific research associated with production;
  * light industrial production which is dependent on regular consultation with the tenant's own research, development and design staff established in the Cambridge area; or the scientific staff of the University or of local scientific institutions;
  * ancillary activities appropriate to a Science Park.

Availability of land

* 50 hectares; buildings from 50 sq metres to 1200 sq metres;
* Sites for development by the occupiers on long ground leases;
* Purpose-built units on 25-year leases;
* Starter units for smaller companies or 'listening posts' for larger companies, on 3/9 year leases.

Text Box 5.1 Prospectus of Cambridge Science Park (Trinity College, 1989). Cambridge Science Park is a good example of a private sector initiative designed to stimulate the spin-off of technological innovations from a world renowned university. Science parks, technology centres and enterprise zones are some of the measures used throughout Europe to promote high technology development and the establishment of new enterprises.

Scenarios

The most important trends in Europe are the decline of traditional manufacturing and the corresponding growth of service activities. The transition to a post-industrial society is accompanied by fundamental changes in the

Figure 5.2 The growing European economy. Already in 1993, the EC and the countries of the European Free-Trade Association (EFTA) will join to become a free-trade region of more than 375 million people, and this will only be a prelude for closer links with other European countries ranging from loose association to full membership.

character of work whereby the traditional division of labour is being superseded by more flexible working arrangements and a growing physical separation of activities within the large corporations.

What will be the impact of these trends on the economy of Europe? The seed scenarios in Scenario Box 5.1 represent three contrasting positions.

The *growth scenario* depicts an economically strong Europe as the

A   *Growth scenario*

The fears that Japan would dominate the world economy turned out to be wrong. By the turn of the century, Europe clearly is Number One. With its 450 million consumers and the variety of skills and talents in its multinational labour force it presents an economic empire of unprecedented magnitude. Much of this success has been due to the boost received when the east European countries converted to market economies in the 1990s, and the ability of major European companies to consolidate and expand their trading activities throughout the whole world after 1993. However, economic growth is unevenly distributed in Europe. The central European countries, which have retained and constantly modernized their manufacturing base, have fared best, while the southern countries have to rely on less profitable tourist and service economies (see *Regional development*). Yet even in the richer countries income disparities have increased due to redistributing tax policies and declining labour union influence (see *Life-styles*).

B   *Equity scenario*

To reduce the economic disparities between the European countries turned out to be easier said than done. A tax compensation scheme (in which tax revenues from the richer countries were to be transferred to the poorer ones) was diluted by national egoism in the European parliament. The Single European Market introduced in 1993 established equal opportunities everywhere in the EC, but also wiped out many formerly protected industries in particular in the more peripheral countries. So the main thrust of government policy has remained compensatory: to cushion the consequences of economic structural change by equalizing social security payments across Europe and providing funds for unemployment programmes and vocational training. More successful are the policies directed at the economic promotion of individual regions within Europe (see *Regional development*).

C   *Environment scenario*

The ecological crises of the 1990s brought about a political landslide. The new European government is primarily concerned with promoting sustainable development. It has imposed taxes on luxury goods and services in order to reduce consumer demand and raised production costs by improving health and safety standards for consumer products and at work places and by enforcing rigorous emission standards for manufacturing and transport (see *Environment*). It has also doubled the amount of money available for protecting environmentally sensitive areas and for the reduction of the use of industrial methods in agriculture. A high proportion of the government R&D budget is now spent on the development of alternative technologies which are environmentally more acceptable and reduce the wastage of non-renewable resources (see *Environment*). One effect of these protectionist policies is the emergence of black markets.

world leader. Because of the size of the home market and the skills of its labour force it is able to compete successfully both with its present rivals such as America and Japan and also with the emerging economies of the developing countries. This scenario is an extrapolation of the perceived benefits associated with the creation of the Single European Market. Such a scenario is characterized by joint ventures between European companies which combine their strengths and exploit the potential opened up by the removal of internal trade barriers. Their efforts are further boosted by an optimistic assessment of the prospects for the transformation of the eastern European countries into market economies.

The *equity scenario* is dominated by internal rather than external factors. It recognizes the conflict of interest that is likely to face politicians who try to override national considerations and reduce regional disparities within Europe. The extent to which revenues can be directly transferred from richer to poorer regions for this purpose is seen in this scenario as rather limited. Consequently it is necessary to rely on compensatory measures such as the provision for unemployment programmes and vocational training. The equity scenario also reflects the conflict between short-term and long-term goals. Its ultimate success is dependent on the extent to which decision-makers are willing to give up some short-term gains in order to strengthen Europe's long-term prospects as a collective entity.

The *environment scenario* envisages a situation where a European government seeks to promote sustainable development following a series of ecological crises in the 1990s. This scenario involves a higher level of government intervention than either of the other scenarios. It assumes increased taxes on consumer goods to reduce demand and stringent measures to limit the use of non-renewable resources. One consequence of measures such as these is likely to be the emergence of black markets and deliberate contraventions of conservation policies. The success of the environment scenario will depend to a very large extent on the degree to which public support can be mobilized to minimize the impacts of these activities.

The views of the experts

To what extent do these three scenarios represent the main options for Europe? In general the experts agreed that the three scenarios covered the range of possible developments. Nevertheless they also added their own qualifications and suggestions particularly with respect to the status of the European economy, the reduction of regional disparities and the nature of sustainable development.

The status of the European economy

Although half the respondents felt that the growth scenario is the most likely scenario for 2020, only a few of them accepted it without some qualification. Some argued, for example, that Europe will not necessarily be the Number One envisaged in the growth scenario, even though the creation of a single market will make Europe a leader in many sectors of the world economy. Other respondents felt that Europe is likely to come up against increasingly strong competition not only from the USA and Japan but also from the emergent developing countries. A Dutch respondent predicted that 'in 30 years from now, new superpowers are emerging, of which Latin America is the most noticeable one followed by Africa, China, and India. International competition will dramatically increase while the real economic power is controlled by some ten multinational companies'.

Several respondents questioned the degree to which the east European countries will have converted to market economies by 2020. As a Dutch correspondent put it 'I do not expect that all the east European economies will have successfully converted to market economies: only some of them'. The potentially destabilizing effect of recent events in eastern Europe was seen by some commentators as having a profound effect on the future of the European economy. One scenario developed by a British respondent suggested that 'Europe basically breaks up as the changes set in motion in the late twentieth century (such as markets being introduced in eastern Europe) make countries like Germany realize that their best interest is to "go it alone". The disparities within Europe will widen considerably although some "surprise countries" do better under the free-for-all than immediate past economic performance suggests'.

Reducing regional disparities

Several respondents pointed out that the equity scenario contains strong pro-growth elements such as the restructuring of industry on a European basis following the removal of barriers. They felt that the combination of growth with equity represents many of the features of existing policies of the European Community which in the long term are likely to be strengthened by the incorporation of the east European states. In the opinion of a French correspondent 'equity as an alternative to growth is a lure. Before considering compensation we have to create added value'.

Some respondents questioned the assumption made in the growth scenario that the central European countries will fare best in the future. In an era of flexibility it was argued that Italy and Spain may experience higher levels of growth than the UK or Germany thereby helping to reduce regional disparities within Europe. A Swedish commentator drew a parallel between the sun belt of the Mediterranean and that of the United States, arguing that the southern European countries may increasingly attract R&D Departments and other skilled labour services because of their favourable climates.

To reduce regional disparities there was general agreement that European governments will have to concentrate their efforts on European rather than national policies. Measures that might be introduced as part of these policies include social security and regional fund transfers across Europe, accelerated infrastructure development in weak regions and directed technological development to support less favoured regions. The main goal of such initiatives, in the eyes of a Portuguese commentator, should be to promote a development phase which reduces the gap between countries like Portugal and the other Community countries and recognizes their natural resource advantages. In addition to these developments, a Spanish respondent felt that it will also be necessary to eliminate some of the main barriers which prevent the entry of these countries into the north European economies.

The nature of sustainable development

The promotion of sustainable development was regarded by many respondents as a major objective in the growth and equity scenarios as much as in the environment scenario. In the eyes of one British respondent this is due to sheer self-interest. 'Even if the environmental scenario in its most rampant version does not come to fruition, significant elements from it will be incorporated into both the growth and equity scenarios. Even the most short-term marketeers will recognize the dangers presented to market expansion by an ecological crisis. Further, advanced technologies for environmental monitoring and control will in themselves provide a major (and eminently exportable) growth sector'. Similarly a French commentator pointed out that 'it is not unthinkable that ecology fosters growth in many industrial sectors with a few losers of course, but these are secondary. Even the car industry can prosper in an ecology driven world'.

Some respondents also argued that the assumption that the environment scenario is low-tech is incorrect within the European context at least. Being user-friendly to the environment is likely to necessitate substantial investment in new technologies. This will require measures to stimulate innovation in cleaner, low-volume, high-tech products. A Dutch correspondent also suggested that a critical look at the role of agriculture in the European Community is necessary to reduce the area reserved for agriculture and increase the amount of land devoted to nature-building activities.

An important measure in implementing any environmental scenario is likely to be taxation. This formed the centre-piece of a scenario developed by an English respondent.

> The main mechanism behind the green market will be taxation on the use of non-renewable resources and on the release of pollutants, in other words, prices upon taking things from or emitting things into the natural environment. Many of the licences auctioned under this system will be transferable between industries, though only to a limited degree between places; the prices will be related to tariffs around Europe (and other major industrial nations) which will function as a second best for a global licensing/pricing scheme to stop the transfer of extraction/emission to the rest of the world. R&D into environment-friendly products and processes will become primarily the responsibility of the private sector with governments monitoring the environment itself and running the licensing system.

Which scenario?

Half the respondents (30) thought that the growth scenario was the most likely scenario for Europe in 2020 subject to the various reservations described above. A substantial minority of respondents also found it impossible to avoid combining elements from the different scenarios in their own vision of the future. Nearly half the respondents (24) preferred the environment scenario for their own countries, arguing as noted above that sustainable development and economic growth are not necessarily incompatible.

Speculation

One of the most interesting features of the responses from the experts is the extent to which they combined elements from all three scenarios in their preferred visions of the future. They recognized that 'significant elements from the [environment scenario] will be incorporated into both the growth and equity scenarios'. Consequently, any speculations about the future must recognize that the three scenarios represent differences in emphasis in terms of policy priorities rather than contrasting or opposed positions.

Consequently the prospects for the future economic growth of Europe are likely to be very much dependent on the extent to which the benefits associated with the Single European Market are realized, irrespective of the preferred scenario. However, the scale of economic growth is likely to be given a higher priority in the case of the growth scenario than in either of the other scenarios. There is plenty of evidence to suggest that the potential gains from European integration may be considerable. The Cecchini Report (1988), for example, estimated that the benefits arising out of the removal of internal trade barriers within the European Community might amount to between 4 and 6 per cent of Europe's gross domestic product (Table 5.2).

The extent to which these benefits can be realized by 2020 and the associated economies of scale exploited by European companies depends on the degree to which cultural as well as trade barriers are eliminated in Europe. There is also the prospect that the creation of a 'fortress Europe' may stimulate aggressive policies from its main international trading partners which will threaten the benefits to be derived from a single market.

Given these circumstances, policy-makers and decision-makers will have to choose between two courses of action:

either   to accept the main features of the growth scenario with its unparalleled wealth and its equally unparalleled congestion and consumption of resources. Such a scenario would lead to greater regional disparities within Europe;

or   to develop some combination of the equity and environmental scenarios which seek to reduce regional disparities and encourage sustainable development. Such a scenario would involve stringent measures to limit resource-consuming activities which would reduce Europe's overall competitiveness in relation to growth-driven economies in other parts of the world.

Given the extent to which the scenarios contain overlapping elements, European policy-makers might also concentrate their attention on three key areas. First, there is the need to overcome the cultural barriers that are likely to inhibit the emergence of a truly single market after the trade barriers are removed in 1992. In this respect it is essential to replace national by European perspectives in investment decision-making and policy formulation and to create and consolidate the networks that are needed for this purpose. Second, some measure of reduction in regional disparities will be needed to exploit the full potential of an integrated Europe. The importance of investment to develop resources in countries such as Portugal and Greece should not be underestimated. Finally, there is an urgent need for investment in alternative technologies that promote

Table 5.2 Estimated economic benefits of the Single European Market

Source of benefits	Estimated benefits	
	Billions ECU	Per cent of GDP
Gains from removal of barriers affecting trade	8–9	0.2–0.3
Gains from removal of barriers affecting overall production	57–71	2.0–2.4
Gains from exploiting economies of scale more fully	61	2.1
Gains from intensified competition reducing business inefficiencies and monopoly profits	0–46	0–1.6
Total for 7 member states at 1985 prices	127–187	4.3–6.4
(Total for 12 member states at 1988 prices	174–258	4.3–6.4)

Source: Cecchini, 1988
Note: The Cecchini Report estimated the costs of non-Europe at more than 200 billion ECU. It argued that these could be converted into potential gains by the removal of existing barriers and the creation of a single market in the twelve European Community countries. These gains are the equivalent of 4.3 to 6.4 per cent of the Community's Gross Domestic Product in 1988.

sustainable development which is generally compatible with economic growth. This will require substantial investment at the governmental level.

Further reading

The issues covered in this chapter are dealt with at greater length and in more detail in a large number of books. However, the six contributions on this topic that are contained in Part I of Pinder's (1990) *Western Europe: Challenge and Change* provide an excellent overview of recent European experience with respect to global competition, corporate restructuring, the emergence of small firms, technological change and producer services. The compilations edited by Williams (1987) and de Jong (1988) also contain a great deal of useful information on recent developments in the European economy.

Technological change and its impacts on regional development is dealt with in an excellent fashion in the compilations edited by Amin and Goddard (1986) and Giaoutzi, Nijkamp and Storey (1988). Another compilation edited by Cooke (1989) contains the findings of a number of detailed case studies of industrial restructuring in Britain. New firms and their consequences for regional development are also dealt with in some depth in the work edited by Keeble and Wever (1986).

There are many works on science and technology parks, enterprise zones and technopolis developments but none of these are specifically European in scale. A good introduction to science and technology parks can be found in Monck et al.'s (1988) survey of British experience, while the volume edited by Smilor et al. (1988) on the technopolis experience contains useful case studies of the 'Cambridge phenomenon' and Sophia Antipolis. The former is also dealt with exhaustively in Segal Quince Wicksteed's (1985) classic study.

The works by Gershuny (1978) and Gershuny and Miles (1983) describe the main features of the transition from the industrial to the service economy and beyond. The geography of the information economy is set out by Hepworth (1989). The impact of multinational firms on the service sector is explored in the compilation edited by Enderwick (1989).

The publication of the Cecchini Report (1988) on the cost of 'non-Europe' led to a great deal of discussion on the future of the European economy. A good overview of the issues involved is contained in the special issue of the *Political Quarterly* which is now available in book form (Crouch and Marquand, 1990).

CHAPTER 6

ENVIRONMENT

Definition

Environmental resources most affected by transport and communications are energy, clean air and land. Indirect environmental effects of transport and communication affect plants and wildlife, water and soil quality. In a wider sense also human life and well-being, safety and protection from noise and physical disturbance and the aesthetic quality of the human environment are environmental resources affected by transport and communications. Environment issues pervade most other chapters of this book, most notably of course those on goods transport (Chapter 9) and passenger travel (Chapter 10). Environmental considerations play a prominent role in the discussion about future policy options in the subsequent chapters on regional development (Chapter 7) and urban and rural form (Chapter 8). Environmental quality strongly interacts, and most often conflicts, with industrial location (Chapter 5), and is a central factor in modern life-styles (Chapter 4).

Trends

About 25 per cent of all final energy consumption is transport-related. In western Europe, like in all industrial regions of the world, total energy consumption per capita has slightly decreased after the energy crises of the 1970s. The differences in energy consumption between the industrialized and developing regions in the world are enormous and pose serious questions of equity which, in conjunction with the 'greenhouse' effect, require immediate action (see Figure 6.1). Even within Europe the differences in energy consumption are considerable with, for instance, Germany consuming more than twice as much energy per capita than Greece or Spain.

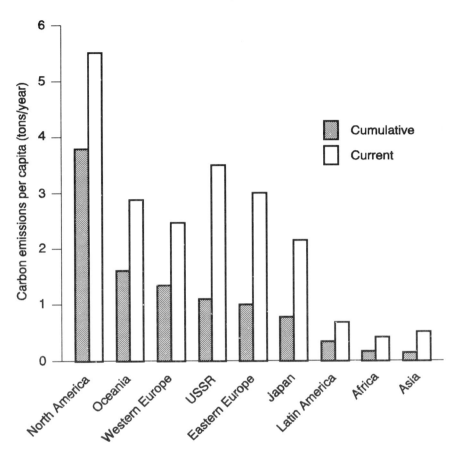

Source: IIASA

Figure 6.1 Carbon emissions per capita by region from the burning of fossil fuels, cumulative (1800–1987) and current (1987). The differences in energy consumption between the industrialized and developing regions in the world are enormous and pose serious questions of equity and, in conjunction with the 'greenhouse' effect, require immediate action. Even within Europe the differences in energy consumption are considerable with, for instance, Germany consuming twice as much energy per capita than Greece or Spain.

Transport-related air pollution is responsible for between 60 and 95 per cent of carbon monoxide (CO) and carbon dioxide (CO_2), between 30 and 60 per cent of nitrogen oxides (NO_x) and nearly all lead emissions. Lean-burn engines, electronic injection, computer-controlled spark ignitions, exhaust circulation systems have over the years reduced emissions per vehicle-km, but in absolute terms automobile emissions have increased due to the growth in the number of cars and the trend to longer trips and higher speeds (see Chapter 10). There is clear evidence that automobile emissions are a health hazard in densely populated areas, contribute to the 'greenhouse' phenomenon and endanger the biosphere, as testified by the dying forests in West Germany (see Figure 6.2).

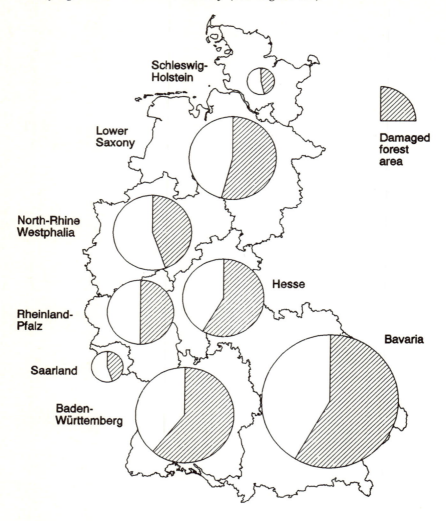

Source: Umweltbundesamt, 1990

Figure 6.2 Proportion of damaged forests in West Germany. There is clear evidence that automobile emissions are a health hazard in densely populated areas, contribute to the 'greenhouse' phenomenon and endanger the biosphere, as testified by the dying forests in West Germany.

In densely populated western Europe, land consumption for transport infrastructure is a serious problem. In agglomerations, railways and roads may occupy one-fifth and more of the total developed area. Large contiguous paved areas prevent rain water from filtering into the ground and are one of the reasons for the lowering of ground water levels; high-speed traffic lanes are barriers for animals and seriously restrict their movement; the fall-out from automobile emissions, in particular dust and lead, contaminate an area about three times as large as the roadway itself. In built-up areas, railways and trunk roads cut through residential neighbourhoods and force pedestrians to make time-consuming detours. Multi-lane freeways and trunk roads disrupt the human scale of historical town centres and their public spaces. The most obnoxious environmental impact of transport, however, seems to be traffic noise. According to OECD estimates, 50 per cent of the population in the United Kingdom were (in 1980) exposed to noise levels of more than 55 dBA outside their residence; the corresponding figures for other countries were 44 per cent for France, 40 per cent for The Netherlands, 38 per cent for Denmark and Sweden and 34 per cent for West Germany.

Finally, automobile transport continues to demand its death toll. In West Germany, for example, more than half a million people have died in road accidents since 1950. It is true that through safer cars and better roads traffic deaths per 100,000 population have more than halved between 1970 and 1985 (from 32 to 13), but this still means 8,000 fatal accidents per year. The European countries also differ widely in this respect, between Greece with 20 people killed in traffic per 100,000 population per year and The Netherlands with only 10 (see Figure 6.3). While the number of fatal road accidents has decreased, accidents causing injuries have increased dramatically. Today, for every person killed on the road there are about 40 persons injured.

The European countries have reacted in various ways to the growing negative impacts of the motorization of society. During the energy crisis, all countries imposed strict general speed limits on highways and, except West Germany, also on motorways. West Germany, however, was the first European country to push seriously the introduction of catalytic converters, much to the displeasure of the car industry and its less environment-conscious neighbours. The three-way catalytic converter, which requires unleaded petrol, is able to clean some 80 per cent of the remaining CO and NO_x, but unlike in the USA and Japan, the adoption of this technology is slow and very uneven in Europe (see Figure 6.4). The Netherlands pioneered a particular combination of car traffic restraint and street design in residential areas, which quickly spread to other countries, most notably Germany, Britain and Scandinavia. In Germany, 'area-wide' car restraint, which includes the scaling-down of trunk roads and extensive speed-limit areas of 30 kmh, are widely applied and accepted by the population. Also

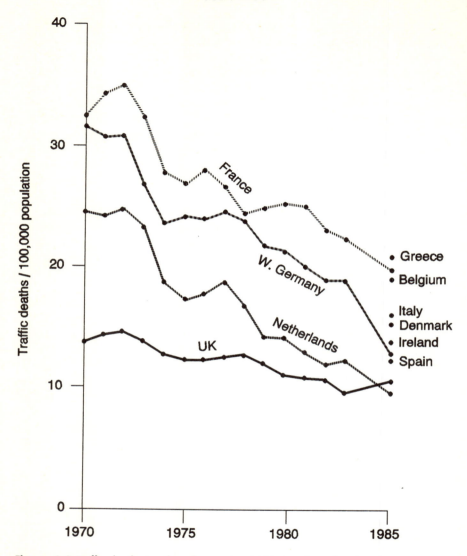

Figure 6.3 Traffic deaths in selected countries, 1970–85. Through safer cars and better roads traffic deaths per 100,000 population have more than halved between 1970 and 1985, but this still means 8,000 fatal accidents per year. The European countries also differ widely in this respect, between Greece, with 20 people killed in traffic per 100,000 population per year, and The Netherlands with only 10.

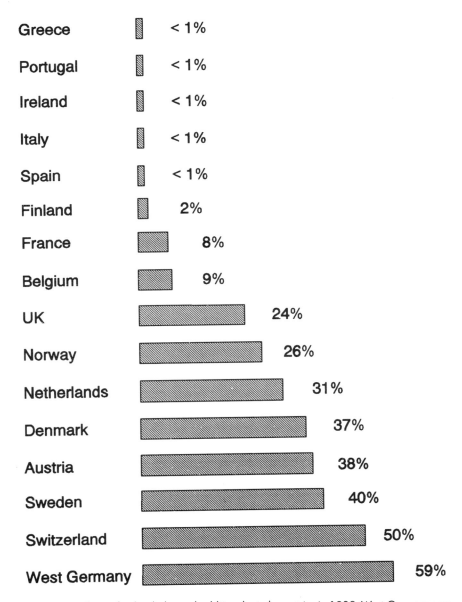

Figure 6.4 Share of unleaded petrol sold in selected countries in 1989. West Germany was the first European country to push seriously the introduction of catalytic converters, much to the displeasure of the car industry in its less environment-conscious neighbours. The three-way catalytic converter, which requires unleaded petrol, is able to clean some 80 per cent of the remaining CO and NO_x, but unlike in the USA and Japan, the adoption of this technology is slow and very uneven in Europe.

in Germany, pedestrianization of inner-city shopping areas has been applied in more than 800 cases. In other countries, the fight about the future role of the automobile in the city is still going on, with Bologna and Florence marking the first victories of a peaceful coexistence between people and cars, while Rome and Athens are still battlefields of resistance of the hardline car front.

The other obvious approach to make urban transport ecologically less harmful is to promote or revitalize public transport. Passenger rail systems and even buses are incomparably less polluting per passenger kilometre than the car (see Figure 6.5). Therefore many cities invest in improvements to their existing systems or in new public transport systems or experiment with new fare systems to attract more passengers (see Chapter 10). The problem is that except in the largest cities the operation of public transport systems without heavy subsidies is impossible. However, in many countries it is beginning to be felt that the car and public transport should no longer be considered as competing but as complementing components of urban transport and that part of the costs for public transport should be financed by the general public just as are the costs for roads.

Wherever large-scale transport infrastructure projects are at stake, there are controversies between environmentalists and engineers. There is hardly any airport extension, new motorway or high-speed rail track which is not fought through all levels of court by some opponents. Examples are the new airports for London or Munich, the *Via do Infante* between Seville and Lisbon or the high-speed trains between Seville and Madrid, Aix-en-Provence and Nice or Hanover and Würzburg to name but a few. These battles sometimes lead to a change of the project, sometimes to its abandonment, but always to a substantial delay. More and more agencies try to circumvent opposition by proposing expensive tunnel solutions—a way out only for the richer countries. People in less affluent countries can hardly afford to be so particular about the environment and still gladly accept any piece of infrastructure they can get.

The overall trend scenario with respect to transport and the environment is therefore partly cheerful and partly gloomy. On the cheerful side, there are certain further advances in transport technology bringing still cleaner, more energy-efficient cars, more sophisticated electronic techniques of traffic control and guidance, probably even *really* clean (hydrogen) fuel. On the more gloomy side, there is the prospect of ever more cars and trucks demanding ever more road space, and hence ever more roads and motorways ploughing through remaining environmental reservoirs of Europe and choking our cities and filling them with never-ending traffic noise.

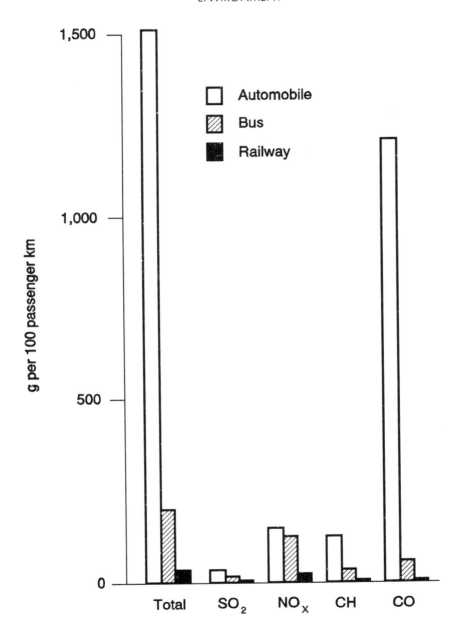

Figure 6.5 Comparison of emissions of car, bus and rail, 1987. Passenger rail systems and even buses are incomparably less polluting per passenger-km than the car. Therefore many cities invest in improvements to their existing or in new public transport systems or experiment with new fare systems to attract more passengers.

Scenarios

To summarize, there has been progress in many areas of environmental protection, but some problems are growing faster than the countermeasures, and many of them are transport-related. On a global scale conflicts are appearing between the industrialized countries, which are the largest users of energy and largest polluters, but can afford to protect their environment, and the developing countries for which economic growth has priority. On a lesser scale such conflicts exist also within Europe.

Given these trends, several policy scenarios are possible. As in the other chapters, they are here subsumed under the political paradigms of *growth, equity* and *environment* (see Scenario Box 6.1). Which of them will be dominant and which of them will be most desirable?

The *growth scenario* is a mixture of high-tech fantasy and ecological nightmare. The 400kmh *SuperTrain* stands for all the things advanced transport technology can achieve: speed, efficiency, probably highly profitable. Not mentioned are the negative sides of high-speed rail transport: high land consumption, energy use, noise. In a deregulated economy only the most successful transport modes survive; the price for the high accessibility of the centres is the increasing isolation of the periphery and the 'grey zones' not served by the high-speed modes. Distant locations become nearer, but near locations more distant. The most adaptable and aggressive transport modes, car and truck, grow fastest, with predictable effects. London, Rome and Athens are examples of possible makeshift responses to the unsolvable problems of a society built on private motorization.

The *equity scenario* looks into the environmental effects of an equalization of economic development in Europe. The scenario explicitly refers to an European infrastructure policy aimed at reducing the differences in accessibility between core and periphery and between city and countryside. One can imagine that it is part of a policy mix containing the full range of regional policy measures designed to assist retarded regions. The price to be paid for this is a decline in international competitiveness. The scenario leaves it open if that is real or just the perspective of the international business community. After all, a more decentralized economic structure might well be more efficient than a highly centralized one suffering from negative externalities of over-agglomeration. Under a decentralized policy environmental quality should become better in urban areas, but may also become worse due to rapid economic development in peripheral regions; the overall balance is still likely to be positive.

The *environment scenario* extrapolates the tendencies already observed in many European countries to reduce radically environmental pollution. The policies applied include international agreements, raising of environ-

A *Growth scenario*

In the year 2020, the 800-km train ride from Paris to Berlin takes less than two hours (see *Passenger transport*). In a deregulated transport market, only the *SuperTrain* survived against the competition of the airplane and automobile; regional passenger and freight service have long been abandoned. As a consequence, the car and truck populations of Europe have doubled since 1990 and so have congestion and pollution on motorways and in urban areas. A completely new network of European toll motorways is nearing completion; some of them cut through the last remaining national parks. The first truck-only motorway was opened in 2003 (see *Goods transport*). Congestion in London is controlled by rigorous road pricing: the City cleared for Rolls Royces. Rome and Athens are drowned under a sea of illegally parked cars and permanent smog; most of their antiquities are permanently damaged.

B *Equity scenario*

It had not been easy to resist the pressure by the national railway companies to close down unprofitable rural services (see *Passenger transport*), but the European Parliament agreed that the only way to reduce the disparities between central and peripheral countries in Europe was to promote decentralization. This meant less money for high-end infrastructure such as airports, high-speed trains and high-capacity telecommunication networks, but more support for the modernization of the rail and highway systems of Portugal, Spain, Greece and Turkey. This policy was only partly successful. It did help to reduce regional disparities within Europe (see *Regional development*), but it also constrained its global competitiveness in the eyes of the international business community. In addition, the policy had the undesired side effect that some of the few remaining natural reservoirs at the Black Sea coast and in the Algarve fell victim to industrial and commercial development.

C *Environment scenario*

By 2010, Europe has become a leader in environment-conscious policy-making. Despite controversies during the 1990s, legislation to clean up the North Sea and the Mediterranean was finally passed. Emission standards for industry and transport are stricter than in the USA and Japan and the use of fossil fuels has been kept constant since 1995. Several large transport projects, such as the *Transrapid*, the second Brussels airport and the St Gotthard base tunnel were abandoned). Heavy taxes on car ownership and petrol brought car ownership down and made public transport almost profitable. Many people, not always voluntarily, moved back into the city; this was good for small cities, but in large metropolitan areas commuting times have become excessive. Fortunately, the introduction of the three-day working week in 2006 made this more acceptable.

mental standards and of taxation of energy consumption and car ownership and use, as well as the abandonment of large infrastructure projects.
Although the scenario does not mention it, the conflict lines are, as in the
equity scenario, between the defenders of this policy and industry, which
is concerned about loss of international competitiveness. Nevertheless,
economic prosperity does not seem to be absent, otherwise the transition
to the three-day work week referred to would not be realistic. The metaphor
raises the issue of qualitative versus quantitative growth. The scenario
forecasts a back-to-the-city movement as a consequence of the restrictions
on car ownership and use but does not fail to indicate the inconvenience of
life in very large cities entirely depending on public transport. So the environment scenario would certainly benefit the environment but might not be
full bliss in every other respect.

The views of the experts

Again it is now asked how the respondents of the survey judged the seed
scenarios. Are these the three most relevant options or are there others?
Are they consistent in themselves? Are combinations of the three scenarios
more plausible? What is missing?

There was a dominant feeling by the experts that the three seed scenarios
were too extreme and that the most likely future would contain elements
of all three or at least two of them. Yet there was little agreement about
the precise mix of these elements. In particular two views can be distinguished: the first cannot see a conflict between growth and environment,
and the second sees a natural link between environment and equity. There
are three more significant views, frequently intertwined with each other or
the first two. One group of respondents put all their faith in a system of
user fees in order to internalize environmental costs. A second group
emphasizes the importance of environment-conscious, integrated land use
and transport planning. A third group thinks that a fundamental shift of
awareness of consumers and producers is necessary to solve our
environmental problems. Finally the last group of comments criticizes the
Eurocentric view of the seed scenarios. These six views will be discussed
in the following paragraphs.

Growth vs. environment

A large group of the respondents feels that it would be wrong to construct
a conflict between economic growth and the environment. They see

numerous ways in which advanced technology can contribute to a better environment and state that every environmental policy can only be financed through growth. Conversely, sustainable growth is not possible without a balanced environment.

Yet above all economic growth depends on efficient transport systems, so airports, high-speed trains and motorways will continue to be built, if necessary against citizen protest, and so will new transport modes such as underground tunnel systems and automatic road guidance systems: 'Despite the increase in mobility and in goods transport, new systems have been able to cope with environmental decay. The most noticeable one is the design and construction of subterranean vacuum pipelines for fast goods transport in standardized containers/pallets. Experiments for passenger transport using the same technology are progressing'. There is a striking agreement on the future importance of underground transport: 'Urban transport will be predominantly underground, which will contribute substantially to a healthier environment in the cities'. One respondent points to the plans for underground toll car networks in Paris and London, while another calls for underground conveyor belts for goods distribution in large urban agglomerations.

That increasingly clean fuels will be developed and used is taken for granted by the majority of the experts; the same applies to battery and solar energy driven cars, even though their introduction will require heavy public subsidies. In addition there is the hope that high-capacity telecommunication networks and 'soft' telecommunication technologies will reduce the need of the business community for 'super-infrastructure': 'The congestion problem will not be solved through road network expansion but rather by telecommunications'.

Equity vs. environment

A number of the respondents found it difficult to choose between the equity and environment scenarios, because they felt that social and ecological aspects cannot be separated. In some respect, they felt, environmental policies even had an equalizing effect. For instance, strict environmental standards in agglomerations should in the long run make more peripheral areas more attractive for the location of firms. It was also questioned whether economic decentralization policies endanger the environment of peripheral regions in all cases. It was stated that at least in one country, Greece, it was possible through a vigorous planning effort to designate the areas for new industries and so fully control their negative effects on the environment.

Regulation vs. user fees

When it comes to policies, there is almost unanimous agreement that market-oriented policies are preferable to regulations. More precisely, everybody seems to agree that strict emission standards with respect to energy consumption, air pollution and traffic noise are necessary and will be implemented in all European countries in the medium term. Also, regulations to restrict the use of certain transport modes in some circumstances, such as the prohibition of short-distance air travel, speed limits for cars or car restraint policies in cities, are considered necessary ingredients of an effective environmental policy. Yet there is no group of measures so unanimously praised as policies built on the user charge principle.

User charges today are a familiar concept in all fields of environmental policy such as industrial emission control, water management or solid waste disposal. However, in transport policy user charges, except of course for public transport fares, are a relatively new issue. In principle, all proposals for user charges are built on the notion that some transport modes, in particular car and lorry traffic, are presently substantially underpriced and would need to become much more expensive if all their negative external effects such as infrastructure cost, environmental damage and costs of accidents had to be taken into account (see Chapter 10). If through appropriate taxation the full social cost could be charged to the user, the resulting tax revenue could be used to subsidize environmentally more desirable transport modes. In addition, and this would be the more important effect, transport users and shippers would be motivated to choose environmentally more desirable modes or reduce their mobility. A third effect would be that the development of less harmful forms of fuel or vehicle technology would be stimulated.

The controversy starts when it comes to the type of user charge and the way of levying it. Some experts propose to tax the purchase of new or used cars (to make car ownership more expensive), some a higher tax on petrol (to make car use more expensive). The majority, however, favour road pricing, i.e. taxes on driving and/or parking in congested areas. One British respondent foresees that 'almost all car taxation is replaced with a common European (electronic) road-pricing system. It allows for demand management at large scale'. A Swedish expert sees road pricing as only one element in a European system of 'emission charges on the opportunity costs for reducing the same amount of pollution elsewhere. Some of these funds are transferred among nations'. The 'user pays' principle would not only apply to transport:

All products have to be recycled which will be the responsibility of the manufacturers. Life cycle emissions have to be written on all products and the producers have to buy emission rights for the life cycle emissions before the product is produced. These emission rights are set for each region and globally. They are lowered each year and can be traded among producers and users. Product liability is extended to include impacts on the environment.

Perhaps the most important advantage of road pricing would be that it can be selectively applied to reduce congestion in inner-city areas. This would, as one respondent put it, 'allow those who gain the most from private vehicle use to do so' and force 'the wealthy who live in suburbs and increasingly in rural areas to pay the full costs of doing so'. The less well-off have to reduce their mobility or use public transport. Public transport, by virtue of fare competition with private modes, would become profitable and provided where most efficient. Using the tax revenue from road pricing, public transport could even be made free. Formerly unprofitable railway lines in rural areas need not be closed down.

Deregulation

A market-oriented environmental policy built on user charges must not be confounded with present tendencies in some countries to deregulate the transport market. There is one particularly critical comment on these tendencies by a British respondent:

I believe that strong environmental pressure will prevent the widespread adoption of a deregulated transport market, simply because it is both wasteful of resources and largely self-defeating in its own terms. The market is simply not the most effective way of coping with transport needs in metropolitan contexts. Particularly then for major cities, a public transport dominated transport policy will be a prerequisite for continued economic vitality. The recent experience of London, which is suffering chronic traffic congestion as a result of a deregulated, anti-subsidy approach to its transport system, clearly highlights the unsustainability of this model. Although . . . vehicle ownership will increase, and although very considerable sums will continue to be invested in improving urban and interurban road networks, I would expect these measures to be complemented by a variety of more environmentally friendly measures; these will certainly include a major emphasis on urban public transport investment, as well as systems to promote better use of existing road networks, such as those to encourage car sharing which have been pioneered in California. All that is required is a societal consensus to emerge over the essential complementarity between public and private systems of transport provision, translated into policy design which recognizes how they may be combined in differing contexts, and when one is more appropriate than the other.

Integrated land use/transport planning

A large number of comments emphasized the importance of a better integration of land use and transport planning either to reduce the need for mobility or to make transport more environment-friendly and socially acceptable.

On the interregional scale these policies try to promote decentralization on all levels of the spatial hierarchy to develop endogenous economic structures that help to reduce distance of transportation between suppliers and customers. However, it is also recognized that decentralization may only be successful if there are also high-speed rail links to the European core regions. Logistic chains and telecommunication infrastructure are other ways suggested to reduce the transport of people and goods (see Chapters 9 and 11).

Efforts to reduce the need for mobility in cities aim at halting or reversing the increasing spatial division of labour and space consumption of all urban activities in favour of a closer association of residences, work places and the locations for shopping and leisure. These policies try to promote more compact, mixed-use forms of settlement in which travel distances are shorter than in today's cities. Also the advantages of different time regimes for cities are discussed. By allowing for individual variations in working time (flexitime) and recreational preferences, cities can be better adapted to public transport and land constraints. Another approach to reduce intra-urban mobility is to give incentives to change locations of living, services and work places instead of commuting.

On the transport side the proposals concentrate on ways to promote environmentally desirable transport modes; public transport in the first place, of course, but also bicycles and walking and various forms of paratransit such as bustaxi hybrids, as well as small airports with more sophisticated control systems. One respondent sees a future in which 'car use in urban areas has gone down and public transport and walking and cycling have intensified due to 30kmh zones, exclusive pedestrianization and improved bus/train and walking/cycling facilities'.

Awareness and values

One group of respondents argued that all environmental policies will fail if they are not accompanied by a fundamental change of awareness and values in the population. One respondent from Germany commented: 'All A scenarios are *laissez-faire* scenarios, whereas social and ecological goals in the B and C scenarios are always to be achieved by policies "from above". This is not realistic. Developments in the direction of the B and C scenarios can only be brought about by a change of values and consciousness in the population. Environmental protection by coercion will not work'.

Some see already a shift in public opinion towards an environmental life-style and 'new' attitudes. However, a structural change of consumption behaviour would be needed, such as a more critical attitude towards new products or a new form of 'soft tourism'. Different forms of consumer behaviour would also have effects on the behaviour of producers, not only with respect to cleaner and more natural products. A trend to less capital intensive technology could mean more labour-intensive production and less pollution.

Which scenario?

In the vote count, the environment field showed the greatest polarization of opinion. About half of the respondents (26) felt that the partly attracting, partly frightening outlook of the growth scenario was the most likely. Only a minority (9) saw a chance for a European transport policy promoting regional decentralization by developing transport networks at the European periphery at the expense of high-end infrastructure as foreseen in the equity scenario. However, a relatively large number of respondents (17) held the optimistic view that environmental policy as described in the environment scenario would become reality.

The environment scenario was the clear winner as the most preferred scenario. More than half of the respondents (34) wished that legislation for clean seas, energy conservation and car constraint would be passed in Europe, and more than two-thirds if those votes are included that favoured some combination between the scenarios A, B and C. Only a very small minority (6) preferred the growth scenario.

Speculation

The most remarkable feature of the responses of the experts in the environment field is their general optimism. There is no feeling of an unavoidable environmental apocalypse, but rather a pragmatic assessment of the risks and challenges lying ahead, probably a result to be expected from a panel consisting of planners, economists and engineers.

The policies suggested cover a wide spectrum from high-growth and high-tech solutions to proposals to abandon the growth-oriented, energy-consuming way of life of Western-style industrial societies. The high-growth/high-tech philosophy is based on the belief that technology properly applied can solve both equity and environment issues associated with industrial growth. It is the philosophy of the rationalist and engineer, and, despite its association with advanced technology, is essentially conservative in anthropological terms. The rationalist imagination is more apt to design a new technology than to anticipate an adjustment in human behaviour. Therefore it takes the genetic disposition of man to expand his territory, consumption and mobility as a given fact and tries to provide the technical solutions needed for this expansion.

Of course the rationalists are also well aware of the eventual clash between unlimited growth and environment in a world of limited resources, but they are convinced that this crash is still far away and that they have the means to keep it away, maybe for ever. So they are very much concerned about possible opposition to their plans by people they think are good-willed but poorly informed, and make great efforts to defend their usefulness. Some of the most frequently advanced positions in recent debates are:

— Environmental policy requires economic growth.
— High mobility and environmental protection are not in conflict.
— Energy conservation can be achieved by more energy-efficient engines.
— Air pollution can be reduced by clean fuels.
— Road transport informatics will reduce road congestion.
— Decentralization requires efficient transport connections.

Each of these assertions can be refuted if a more comprehensive system view taking into account context variables and feedbacks is applied:

— It is true that affluent societies can better afford to invest in environmental measures. However, if all countries would follow the growth path of the industrial nations, the world's resources would quickly be exhausted.

— The damage to the ozone layer by high-flying airplanes, the noise intrusion of trains and motorways in densely-settled areas and the energy necessary to move large masses will set a permanent limit to human mobility.
— All potential savings due to more energy-efficient engines have in the past been absorbed by larger cars and more and longer trips.
— So far no breakthrough in really clean fuels is in sight. Battery-driven engines are clean in operation but produce more CO_2 (at the power station) than petrol-driven cars. Hydrogen, unless generated by water or solar power, produces substantially more. Bio-fuels would be neutral with respect to CO_2 but would require a fundamental restructuring of agriculture.
— The increase of road capacity by improved informatics is estimated to be in the range of 10 per cent and so will be quickly eradicated by the general growth in traffic.
— On a European scale, it is very much debated whether improved transport links between central and peripheral regions will help the periphery to catch up with the centre or reinforce the polarization between centre and periphery. On the urban scale, efficient transport induces dispersal but no self-contained decentralization; decentralization with a close association between residences and work places requires that transport becomes slower or more expensive.

These counter-arguments have not prevented the above assertions being used over and over again. Because they fit so well into the strategies of powerful industrial interests, it can be expected that they will continue to play an important role in the debate about the future of transport in Europe.

In passing, a note on tunnels. It is interesting to note how many times tunnel solutions are proposed in the responses. Tunnels as such, except tunnels under sea channels, rivers or mountains, have no genuine advantage over surface connections: they are much more expensive and require complex ramps and exhaust and security installations. In the proposed applications, their only advantage is that they avoid collision with conflicting land uses and with economic, political and social conflicts on the surface. Tunnels are the epitome of conflict avoidance in rich societies.

The other extreme position is that of the radical environmentalists. They believe that without a fundamental change in human behaviour ecological crises are unavoidable. Pragmatic environmentalists believe that, too, and support any policy that promises to influence behaviour in the right direction, but at the same time appreciate any ecological advance in transport technology. They meet with pragmatic rationalists who realize that without some constraint on mobility even the most sophisticated

high-tech solutions do not work. This is the broad coalition for road pricing. Most likely road pricing is indeed the only really effective policy to
control both congestion and pollution. It is surprising, though, that almost
none of the respondents seemed to be concerned about the serious equity
and privacy risks associated with electronic road pricing schemes. While it
seems likely that the privacy issues might be solved by even more sophisticated electronic safeguards, the equity problems of road pricing have not
yet been adequately discussed. If personal automobility is one of the con-
stituent freedoms of modern life, any rationing or pricing system must be
connected with some kind of deprivation.

Faced with these options, European and national policy-makers will
have to choose between two courses:

either   to continue to pay lip-service to ecological concerns but also to
         make concessions to growth interests of countries and power-
         ful industry groups; to settle down with the lowest common
         denominator between the European countries where environmen-
         tal standards are concerned; and to continue to promote one-
         sidedly high-speed transport infrastructure for the sake of global
         economic competitiveness and efficiency;
or   to join the few enlightened industry leaders who realize that for
         environmental and global equity reasons the aggressive growth of
         the industrialized countries must be slowed down; to orient Euro-
         pean environmental policy not at the level of environmental pro-
         tection of the least but of the most advanced nations; to redirect
         transport investments to improving environmentally desirable
         transport modes and transport infrastructure in peripheral
         regions.

In reality, the choice may not be so black or white. In most European
countries the time for a farewell to growth-oriented economic policies may
not yet have come. The Netherlands is a country where a persuasive compromise between the requirements of a growth-oriented economy and
environmental protection has been found. The result is clearly a mix of
policies relying equally on technological innovation as on influencing
behaviour. Figure 6.6 shows, as an example, the expected impacts of
policies to reduce the emission of nitrogen oxides from private cars and
heavy goods vehicles. Through a combination of technical, fiscal and
demand management measures a reduction of transport-generated NO_X
emissions to a quarter of current levels is possible without substantial
restrictions on mobility.

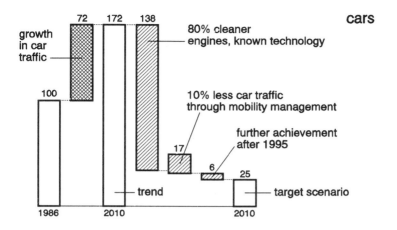

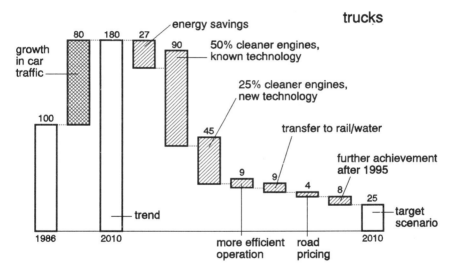

Source: Ministry of Transport and Public Works, 1989

Figure 6.6 Effects of policies to reduce the emission of nitrogen oxides from private cars (top) and heavy goods vehicles (bottom). Through a combination of measures a reduction of transport-generated NO_x emissions to a quarter of current levels is possible without substantial restrictions on mobility.

Further reading

During the last two decades, the urgency of global problems posed by continuing population growth, depletion of resources and climatic change has been brought into public awareness by documents such as the Club of Rome's *The Limits to Growth* (Meadows et al., 1972), the *Global 2000 Report to the President* (Barney, 1980) and the *Report of the Brundtland Committee* (World Commission on Environment and Development, 1987), and has only recently been dramatically underlined by the latest report of the Club of Rome (King and Schneider, 1991). Today there is an overwhelming flood of literature on environmental issues on both popular and scientific levels. A comprehensive overview on a popular level is the *The atlas of future worlds* (Myers, 1990), a highly imaginative account of the global interconnectedness of environmental problems. On a more scientific level, Pearce et al. (1989) and Owens and Owens (1991) provide good overviews of concepts, problems, policy approaches and analysis methods in environmental planning. Rees (1990) deals with the issue of scarcity of resources. Barde and Pearce (1990) give an overview of methods to evaluate environmental effects. Johnson and Corcelle (1989) present an interesting overview of the environmental policy of the European community before the much debated *Green Paper* of the Commission (Commission of the European Communities, 1990).

Literature on environmental issues in *transport* is also rapidly increasing. A good account of the issues and problems in this field is contained in *Transport and the Environment*, a volume edited by the Organization for Economic Cooperation and Development (OECD, 1988a). In a companion volume (OECD, 1988b), transport-related environmental problems and policies in ten large metropolitan areas in OECD countries are compared. A third OECD publication (OECD, 1986b) explicitly addresses the problem of transport noise. Button and Barde (1990) present six case studies about transport and the environment in six European countries. Northcott el al. provide good information on greenhouse gases (1991).

The Greening of Urban Transport (Tolley, 1990) is a topical collection of papers dealing with new tendencies in transport planning in European and American cities, such as pedestrianization, traffic calming and other measures to contain the automobile and promote 'soft' transport modes such as walking and cycling. In this field, the British literature is in a process of catching up with continental countries such as The Netherlands or Germany in which the critique of the dominance of the automobile and its negative impacts has a significantly longer tradition. Of the many contributions of this kind, Vester's (1990) *Ausfahrt Zukunft* (*Exit to the Future*) deserves mention here despite its being written in German. In his book Vester treats the car as an element of a system consisting of man, economy and the biosphere and demonstrates that only a radically new approach can overcome the fatal car-dependency of modern industrialized societies.

CHAPTER 7

REGIONAL DEVELOPMENT

Definition

These component scenarios deal with the processes of economic growth and decline in central and peripheral regions of Europe and the spatial disparities resulting from them. Particular attention is paid to the role of transport and communication infrastructure for regional development. Though treated in a separate chapter here, regional development cannot be seen in isolation from the spatial developments *within* the regions. These will be dealt with under the heading of urban and rural form in the following chapter (Chapter 8). Similar close links exist between regional development and goods transport (Chapter 9) and passenger transport (Chapter 10) and, more recently, communications (Chapter 11). The links to population (Chapter 3), through migration, and economy (Chapter 5), through industrial location, are obvious.

Trends

After thirty years of European regional policy (in the EC), the differences in income, employment, infrastructure and provision of services between the regions of Europe are still enormous. If the mean gross regional product of the EC is taken to be 100, regions vary between 218 (Hamburg) or 175 (Paris) and less than 50 in parts of Greece and Spain and less than 30 in most of Portugal (1984). These numbers are only indicators of other inequalities in economic opportunity and quality of life. Figure 7.1 shows the areas in the EC eligible for Community assistance under the three principal objectives of EC regional policy, whereas Figure 7.2 shows the spatial distribution of the *synthetic index*, a measure of regional economic success integrating both economic strength and unemployment.

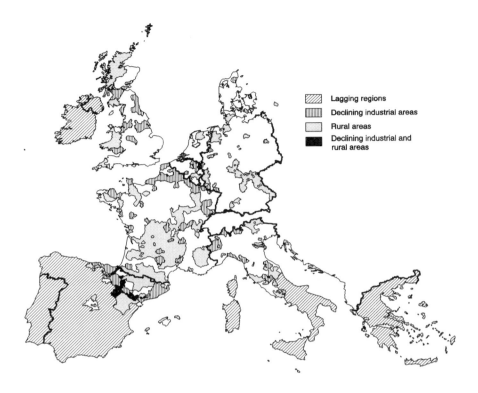

Lagging regions
Declining industrial areas
Rural areas
Declining industrial and
rural areas

Source: Commission of the European Communities, 1991a

Figure 7.1 Areas eligible for Community assistance under the three principal objectives of EC regional policy. Regions lagging behind as measured by GDP per head (Objective 1), declining industrial areas (Objective 2) and rural areas (Objective 5b). By restricting its assistance to the achievement of a limited set of objectives, the Community concentrates on addressing the most serious regional problems.

The differences in income and employment between the regions in Europe can be linked to four origins. First, centrally located regions tend to be more industrialized and economically more successful than the predominantly agricultural regions at the European periphery. Figure 7.3 shows contours of a potential-based centrality index. Two high-accessibility corridors, one from south-east England through the Benelux countries along the Rhine valley to Lombardy, the second from Paris through northwest Germany to Scandinavia, stand out. With few exceptions (e.g. east Bavaria), the top twenty most productive regions in the European Community are located in or adjacent to these corridors, which are served by

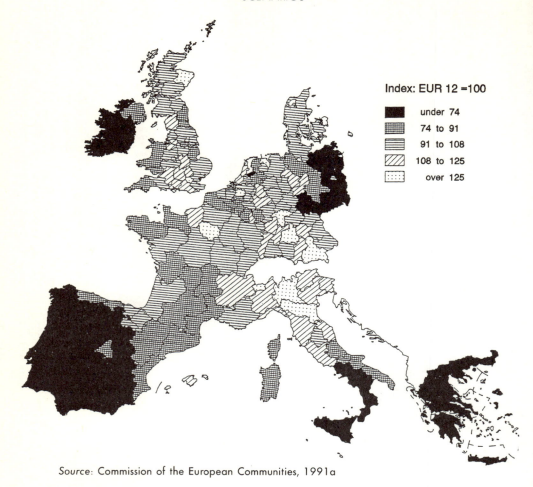

Source: Commission of the European Communities, 1991a

Figure 7.2 Regional disparities in Europe, 1991. After thirty years of European regional policy (in the EC), the differences in income, employment, infrastructure and provision of services between the regions of Europe are still enormous. If the mean gross regional product of the EC is taken to be 100, regions vary between 218 (Hamburg) or 175 (Paris) and less than 50 in parts of Greece and Spain and less than 30 in most of Portugal.

the highest-level ground transport infrastructure and contain the six largest airports in Europe.

Second, urban regions continue to be more productive and affluent than rural regions. Just as in the early days of industrialization, cities are the incubators of innovation where capital and specialized skills come together to produce new technologies and products. Today the hierarchy of cities has become more complex, with a few world cities competing on a global scale, a number of financial and cultural centres of European

Source: Keeble et al., 1986

Figure 7.3 Centrality index contours in Europe. Centrally located regions tend to be more industrialized and economically more successful than the predominantly agricultural regions at the European periphery. The contours of a potential-based centrality index show two high-accessibility corridors, the one from south-west England through Benelux along the Rhine valley to Lombardy; the second from Paris through north-west Germany to Scandinavia. With few exceptions (e.g. east Bavaria), the top twenty most productive regions in the European Community are located in or adjacent to these corridors, which are served by the highest-level ground transport infrastructure and contain the six largest airports in Europe.

importance and a very large number of 'ordinary' cities (see Chapter 8). Not surprisingly, all of the first and most of the second category are within the two corridors mentioned above, while many peripheral regions have a lack of urban centres.

A third factor which effects the economic success of a region is its industrial heritage. Regions that led the first cycle of industrialization based on raw materials and heavy industry such as north-west England, Nord-Pas-de-Calais in France and the Saar and Ruhr regions in Germany suffer from the decline of their major industries (see Chapter 5). Due to their outdated infrastructure, polluted environment and lack of urban amenities, they fail to attract enough of the growing high-tech manufacturing and information-based service industries to compensate for the decline in traditional employment. These new industries select regions with quite different locational characteristics such as high-skill labour, access to international transport and communications networks, availability of business-oriented services and attractiveness in terms of housing, education, culture and recreation.

The fourth factor is the growing dependency of regional development on world markets. As production technologies develop and wage levels in Europe rise, low-skill manufacturing processes may have to be exported to low-wage overseas regions, but may return in the form of the fully-automated factory. Even within Europe labour cost differentials between countries induce a growing internationalization of economic activities in the form of separation of high-skill and low-skill processes within multinational corporations (see Chapter 5). Each region has to develop its material and human resources to find a niche in this rapidly changing global environment. Unfortunately, again the core regions command the information and skills necessary to respond to these international challenges. The Single European Market will remove many of the existing protective trade barriers and so increase the exposure of the European regions to international competition.

The spatial trends resulting from the interaction of these four factors can be summarized as follows. Regions in the core of Europe will benefit most from the ongoing economic restructuring, except the old industrial regions which will continue to decline. Peripheral regions will grow, but less than the core regions, which means that spatial disparities in Europe will increase, although on a higher level.

Current trends in transport and communications are likely to reinforce rather than counteract these tendencies. There is still a vast discrepancy between the transport infrastructure in core and peripheral countries (see Table 7.1). Yet the planned high-speed rail links and new motorways largely follow present transport corridors and so further improve the accessibility between the core regions at the expense of the periphery. Deregulation of air transport will enhance the position of the present major airports serving as hubs routing flights to secondary airports and concentrating traffic in the core regions. Advanced telecommunication services will be available first where high-volume demand will make them profitable, and this will again be in the existing centres (see Chapter 11).

The most likely scenario of regional development in Europe therefore looks very much like the map in Figure 7.4. There will be a wide curved zone of intensive development stretching from south-east England across the Channel (via the Channel Tunnel) through Benelux, the south-west of

Table 7.1 Transport infrastructure in European countries, 1985/86

Country	Railways m/m²	Motorways m/m²	Other roads m/m²
Austria	69	15	1,286
Belgium	120	50	4,312
Denmark	57	14	1,615
Finland	18	<1	225
France	64	11	1,470
Germany, F.R.	110	34	1,942
Greece	18	<1	306
Ireland	28	<1	1,340
Israel	42	10	607
Italy	54	20	966
Netherlands	68	48	2,321
Norway	13	<1	258
Portugal	39	2	570
Spain	25	4	331
Sweden	26	3	299
Switzerland	123	24	1,710
Turkey	11	<1	297
United Kingdom	69	12	1,522
Yugoslavia	37	4	446

Source: Eurostat, 1988
Note: There is still a vast discrepancy between the transport infrastructure in core and peripheral countries. Yet the planned high-speed rail links and new motorways will largely follow present transport corridors and so further improve the accessibility between the core regions at the expense of the periphery.

Germany and Switzerland to Lombardy (the 'Blue Banana'). There it meets with another growth zone developing along the Mediterranean down to Barcelona and Valencia (the 'Sunbelt'). With the exception of the Île de France, even established economic centres outside these two zones will be at a relative disadvantage. However, the opening of the borders to eastern Europe is likely to provide additional opportunities to outlying centres such as Hamburg, Berlin and Vienna.

Scenarios

To summarize, all signs point to a more polarized rather than a more equalized regional development in Europe. Increasing differences in accessibility, continuing urbanization, the growing importance of high-tech industries and the internationalization of markets, taken together,

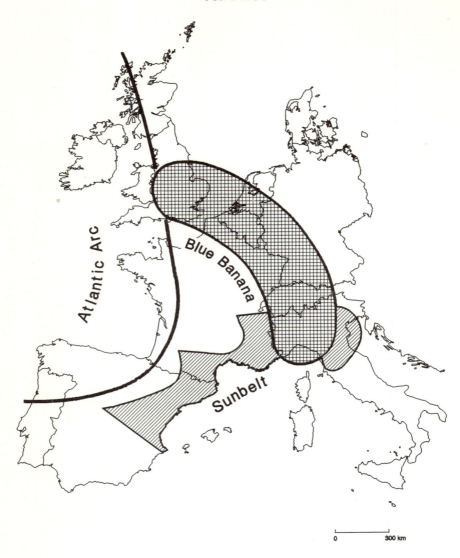

Source: After RECLUS, 1989

Figure 7.4 The 'Blue Banana'. The most likely scenario of regional development in Europe looks like this map. There will be a wide curved zone of intense development stretching from south-east England across the Channel (via the Channel Tunnel) through Benelux, the south-west of Germany and Switzerland to Lombardy (the 'Blue Banana'). There it meets with another growth zone developing along the Mediterranean down to Barcelona and Valencia (the 'Sunbelt'). With the exception of the Île de France, even established economic centres outside these two zones will be at a relative disadvantage.

A *Growth scenario*

The warnings that the Single European Market would result in a further spatial concentration of economic activities have proved to be correct (see *Economy*). The high-accessibility belt from London to Milan has become a veritable megalopolis with a population of eighty million. The Channel Tunnel, the integrated European high-speed trains and the Gotthard base tunnel (see *Goods transport*) have increased the coherence and domination of this megalopolis, which has already in the 1990s, due to the booming east European economies, spread out its tentacles as far as Berlin and Vienna. The negative side effects are severe agglomeration diseconomies in its centres, such as exploding land prices, congestion and environmental deterioration, while the peripheral regions suffer from economic decline and depopulation (see *Population*).

B *Equity scenario*

Soon after the introduction of the Single European Market, the European parliament realized that without a strong decentralization policy the spatial disparities within Europe would become even greater. The result was the reform of the *European Regional Development Fund* which was given extensive powers to redirect economic activities from the agglomerations to the peripheral areas. Investment programmes such as the *Technopolis Network*, the *Remote Area Highway Programme* and the *Regional Airport Scheme* were coupled with strict land-use control in urban areas, tax incentives for location in non-metropolitan regions and flat rates in long-distance telecommunication services. While some of these policies were eroded by national non-cooperation, in general they were successful, as witnessed by the new high-tech industries in traditionally agricultural regions like Macedonia or Galicia.

C *Environment scenario*

The rigorous policies to force industry and consumers to adopt environment-conscious behaviour (see *Environment*) had far-reaching consequences for the European regions. Some regional industries such as the potash and lignite industries in east Europe have altogether disappeared, others had to reorganize their production fundamentally to meet emission standards. Transport in general has become cleaner, but also more expensive (see *Environment*). Together with strict containment of urban areas, this has discouraged decentralization of economic activities. Accordingly, extensive support programmes to maintain the ecological balance in declining agricultural remote regions have become necessary. In summary, although some competitiveness in economic terms may have been lost (see *Environment*), most people feel that the benefits of ecologically balanced growth outweigh the small loss of material wealth.

tend to reinforce the economic position of European core regions at the expense of those at the European periphery.

In this situation, what are the policy scenarios? Scenario Box 7.1 contains the three seed scenarios presented to the respondents of the survey, as before associated with the three paradigms *growth, equity* and *environment*. Are these the three most relevant scenarios, and which of them is the most likely and which the most desirable?

The *growth scenario* is the extreme intensification of the polarization tendencies observed today. All things come together: the increased competition of the Single European Market and the extra economic push received from the expanding east European markets, the high-speed trains and the closure of 'missing links' such as the Channel Tunnel or the Alpine crossings. The scenario deliberately seems to simplify the economic geography of Europe: What about the declining old industrial areas in the heart of the 'Blue Banana'? And where is the fast growing 'Sunbelt'? However, the negative sides of the polarization process are painted out.

The *equity scenario* is very interventionist. It essentially is a list of all kinds of possible Community programmes to counteract the growing regional disparities due to economic integration and deregulation. The most important of these is the transformation of the European Regional Development Fund (ERDF) from an instrument for allocating subsidies into a powerful supranational agency for investment control. One might say that this is quite alien to the principles of the Single European Market, so it may require some imagination to see it work. The other policies are the standard catalogue of regional policies—some of them are copies of government programmes implemented elsewhere, for instance in Japan. The scenario assumes that in general these policies are successful.

The *environment scenario* assumes that a trade-off has to be made between environmental protection and regional economic growth, but that the loss of material wealth is small and well worth the gain in quality of life. Outdated production in eastern Europe is most affected, as obviously modern industries in western Europe are able to cope with stricter environmental standards. In spatial terms, the scenario assumes that a rigorous environmental policy would reinforce polarization because transport becomes more expensive. This makes support measures for peripheral regions necessary, but, unlike in the equity scenario, these do not aim at economic growth but at maintaining the ecological balance in declining agricultural regions.

The views of the experts

How did the experts respond to the three seed scenarios? In no other field did the seed scenarios turn out to be so controversial. The responses of the experts showed that there exists a great degree of disagreement about the future development of the regions in Europe and the possibilities of influencing it by Community policies.

The comments focused around five major topics. The first one was the centralization/decentralization issue. Many respondents felt that all three scenarios gave a too simplified picture of the dominant spatial trends among the regions of Europe. Another group of comments addressed the issue of regional specialization in the face of more and more internationalized markets. Several respondents emphasized the role of urban regions in Europe and how they influence the balance among the regions. There was more consensus with respect to the future of regional disparities and the need for Community action to reduce them. Finally, again, there were some general comments questioning the concentration on European issues and thereby neglecting the global interconnectedness of the regions of Europe.

Centralization/decentralization

A major criticism voiced by several respondents was that the polarization trend predicted in the growth and environment scenarios, and implicitly also underlying the equity scenario, is a gross simplification of a much more diversified pattern of spatial development in Europe. One respondent argued that there will be a growing spatial separation between knowledge-intensive industries, which will concentrate in the core regions, and other industries with routine operations able to exist also in peripheral regions. So on the one hand, one has to observe a further spatial concentration, while on the other hand, endogenous economies will arise.

Several respondents argued that the growth in the Mediterranean 'Sunbelt' was not represented in the scenarios. One of them wrote: 'Nothing will change macrogeographical trends favourable to the Mediterranean "arc" from Valencia to Milan and Genoa: Spain is likely to be the most powerful country of Europe in the twenty-first century'.

A third group of comments added a different perspective for peripheral regions in Europe. It was argued that 'Small can be beautiful, and the far

West or far East can be more dynamic than the old historical centres in huge cities. Even the centre can move'. It was also pointed out that the periphery is not the periphery in absolute terms and that culture-bound factors influence the flexibility of adaptation of different national, regional and local economies. The Nordic countries, for instance, might be highly able to compete even in the context of the growth scenario, though their internal spatial structure would have to become more concentrated.

Finally it was questioned whether the environment scenario needs to be associated with centralization. The alternative would be a 'green' decentralization based on endogenous regional economies of small, self-supply local networks. There is also the expectation that more environment-friendly cars together with clean fuel could make it possible to combine more dispersed living and increased travel with improved environmental quality.

Regional specialization

Some comments stated that an important consequence of the Single European Market which is not mentioned in the scenarios will be the need for specialization in European regions. Some examples for specialized regional functions are: agriculture in Spain, Portugal, southern France, southern Italy, and the Balkan countries; tourism in Spain, France, Italy and Greece; traditional industries in northern England, east Germany, Poland and Czechoslovakia; high-tech industries and services in the 'Blue Banana' and in some scattered centres elsewhere; natural conservation areas in the Atlantic regions, in northern Europe and in the Alpine countries. Regional specialization also offers opportunities for peripheral regions. Tourism, second homes and conference activities might become their 'base industries', provided they have regional airports and efficient links to the European core regions.

The role of cities

Some respondents remarked that today the distinction between urban and rural regions is more important than the one between core and periphery, and that the successful cities in Europe are rather more dispersed than the megalopolis scenario of the growth scenario suggests. One respondent writes:

In each case, the cities which are doing well are those that have successfully managed to position or reposition themselves within an emerging urban system which, like the economic forces which underpin it, bears little relationship to conventional nation state boundaries. While many or most of these cities are indeed within the European core, a number are not. Within the UK context, for example, commentators are increasingly recognizing the relative economic vitality of the city regions of Glasgow, Manchester and Leeds. Although these areas have undoubtedly benefited from some relocation of economic activities of the congested South-East of England, at another level they are successfully positioning themselves as inward investment locations within the European, rather than UK, economic context. One can perhaps point to the relative economic buoyancy of Barcelona/Catalonia as a similar development. Its success cannot be attributed to its location in the 'core', nor, I would suggest, to its location in the 'Sunbelt'. The explanation is both much more complex, and hence much more interesting, than location *per se*.

Some respondents pointed out that if urban regions are the motors of change, this would have implications for Community regional policy. European decentralization policies would have to be oriented towards offering incentives for the development of smaller and medium-sized cities in peripheral regions. The growth of larger urban agglomerations would have to be constrained by strong environmental and social policies including, among others, quotas for immigration into large urban areas. New European institutions would have to be located in the European periphery, and no more European money should go into the 'Blue Banana'.

Interregional disparities

A much disputed issue was, of course, whether regional disparities in the future European Community will increase or decrease. Most respondents agreed with the growth scenario that without additional Community policies market-driven growth will lead to greater disparities. Only a minority believed that market forces will reduce inequality. Several commented that spatial disparities may be felt more strongly between urban and rural or between environmentally more and less attractive regions rather than between core and periphery. One respondent suggested the future European 'regional development map' will look more like a patch-work quilt than any of the scenarios outlined.

Community policies to reduce interregional disparities therefore, besides policies like the ones suggested in the equity scenario, sl.ould be more explicitly focused on cities and encompass 'softer' infrastructures, such as cultural, educational and scientific besides the more conventional 'hard' infrastructure of highways, airports, etc.

Eurocentrism

Finally it was argued that the seed scenarios were written from a narrow central European perspective. Future deliberations should take account of the relationships, interactions and impacts that the individual scenarios will have on the non-European world. Can the globalization of environmental damage (ozone layer, increasing pollution of the oceans, global warming, tropical forests) be ignored? Will the further integration of the countries of Europe, in particular the Single European Market, create increasing tensions with east European neighbours who are not, or not yet, included? What will be the political and economic consequences of the integration of more countries into the Community? Will the growing contrast between the wealth of the countries of the Community and the poverty of their east European, Mediterranean and north African neighbours lead to large flows of illegal immigration, political conflicts or even wars? (see Chapter 3). It was argued that there cannot be a sustainable European regional policy that does not seriously take account of the interrelatedness of the continent with the problems of other countries in this region of the world.

Which scenario?

Despite the disagreement about the pattern of growth, there was much agreement between the experts that growth will result in further spatial polarization in Europe. Nearly two-thirds of the respondents (39) believed that the gap between the rich and the poor regions in Europe will increase, and only 14 thought that policies to reduce regional disparities will be at least partially successful. Only a minority (6) was happy with this likely development; the overwhelming majority feel that a further polarization of regional development in Europe would be detrimental. Interestingly, the preferences were almost evenly divided between the equity (24) and environment (22) scenarios. This indicates that in this field ecological and equity goals are partly in accordance and partly in conflict. On the one hand, it is clear that decentralization would both benefit the peripheral regions and reduce environmental pressure in the core regions. On the other hand, the rapid development of peripheral regions could have a price in terms of new sources of pollution and the loss of formerly undisturbed natural areas.

Speculation

The analysis of the expert responses has shown that issues of regional development in Europe are extremely complex and controversial. There is little agreement about the likely future pattern of growth or about the impact and effectiveness of possible policies. However, the responses of the experts contributed some important information missing or underrepresented in the trend description and the scenarios. One example is the remarkable growth observed in the 1980s in Spain compared with other Mediterranean regions. There is clear evidence that the regions in Spain are rapidly growing in line with the above average development of the Spanish economy, whereas regions in southern Italy, Corsica, Portugal and Greece are tending to fall further back. There is also little progress in Ireland and Northern Ireland (see Figure 7.5). Other commentators rightly observed that the pattern of regional growth in Europe is much more complex than the simple dichotomy of core and periphery suggests. It becomes particularly clear that the system of regions in Europe is superimposed on a system of urban regions, which have more in common than is explained by their location. For instance, an urban region in the European core may be more different from its hinterland than from other cities in the European periphery.

Beyond these differences in interpretation, there is a considerable degree of consensus about the relationship between growth and spatial polarization. If spatial polarization is defined as an increase in spatial disparities (in terms of productivity, employment, income and other indicators of quality of life), there seems to be agreement that without strong equalization policies market-driven economic growth will give rise to regional polarization, (i.e. a widening of the gap between the prosperous and lagging regions), in relative terms, of course, because in an expanding economy every region may grow, only the prosperous regions tend to grow faster.

It is not so clear whether economic growth leads to environmental degradation or improvement. Obviously a retarded rural region is likely to have less pollution than a heavily industrialized one. On the other hand, only very affluent regions have the means to finance advanced

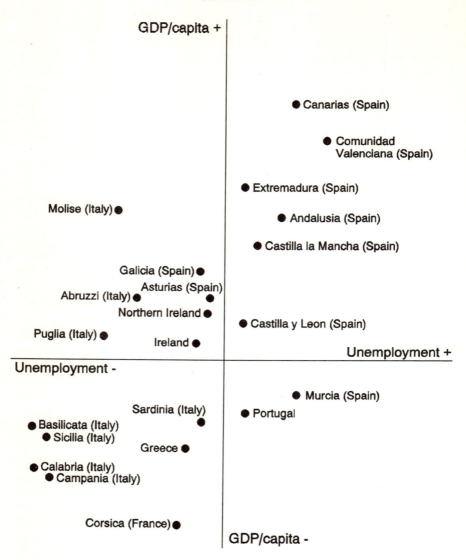

Source: Commission of the European Communities, 1991a

Figure 7.5 Change in the position of the Objective 1 regions compared to the Community average during the 1980s. There is clear evidence that the regions in Spain are rapidly growing in line with the above average development of the Spanish economy, whereas regions in southern Italy, Corsica, Portugal and Greece tend to fall further back. There is little progress in Ireland and Northern Ireland.

environmental protection technology. There is reason to assume that in the initial period of rapid industrial development of a region environmental pollution tends to grow faster than the resources available for environmental protection, so it is in this period that Community assistance for environmental protection is most needed.

So the choices open to decision-makers concerned with regional development in Europe are again related to the issue of economic growth vs. equity:

either   to let capital and economic activity continue to move to the already prosperous central regions of Europe without adequately paying for the environmental damage and agglomeration diseconomies they inflict on themselves and others and to continue to promote the concentration of high-speed transport and high-level telecommunication infrastructure in the already overcrowded European agglomeration and air corridors;

or   to promote a truly decentralized system of European regions with a high degree of regional autonomy enabling them to develop their endogenous potential and narrow the gap between rich and poor regions; to promote the deconcentration of agglomerations to reduce congestion and other agglomeration diseconomies; to promote the transformation of European transport networks towards a dispersed, medium-speed but high-efficiency network without peak loads in space or time.

If one reads the regional policy documents of the European Community, one gets the feeling that the Commission still thinks that it is possible to have both, economic growth *and* the reduction of disparities between the regions. How else is it possible that it simultaneously promotes the expansion of high-tech industries, telecommunication technology and high-speed transport as well as programmes to fight underdevelopment in peripheral regions? A look at the organizational chart of the Commission tells something about the distribution of power and interests in the Community. More than twenty general directorates promote economic growth and try to make Europe more competitive in the global arena. Only one of them (DG XVI) takes care of the losers, i.e. the regions unable to keep up with the accelerating pace of economic competition. The discussion about the value of economic growth if it has to be paid for by deprivation of a large part of the population will go on and it will be exacerbated by the widening gap between Europe and its less affluent neighbours.

Further reading

The literature on regional economic development shows a rapidly expanding field of scientific inquiry with respect to both developing and industrial countries. Vanhove and Klaassen's *Regional Policy: a European Approach* (1987) is probably still the best and most comprehensive book on regional development in Europe, covering theoretical and methodological as well as policy questions. Albrechts *et al.* (1989) is an interesting collection of papers divided into three sections: current dynamics of regional development, policy and implementation issues, and the international-national-local interplay. Pinder's (1990) *Western Europe: Challenge and Change* serves as a good introduction into the major processes at work.

Other publications address specific issues in regional development. Hall and Hay (1980) and Cheshire and Hay (1989) present the results of two comprehensive data collection efforts analysing the growth and decline of urban regions in Europe. They identified the wave-like expansion of urban decline spreading from the 'old industrial' regions in north-western Europe to the still growing urban regions in the south of the continent. Recent work on regional economic development focuses on a small list of recurring themes. Following a recent shift of paradigm in regional economic policy, research concentrates on topics such as 'endogenous' regional potential and the role of innovation and high-tech and telecommunications. Stöhr (1990) illustrates the growing importance of regional self-organization and bottom-up planning relying on resources and skills available but underutilized in the region. Scott and Storper's (1986) *Production, Work, Territory* describes the new geography of industrial development built on internationalization, highly automated production processes and integrated logistics systems (see also Chapter 9). Ewers and Allesch (1990) and Nijkamp (1990) are good collections of contributions circling on the role of innovation for regional development. Hall and Markusen (1985), Blackburn *et al.* (1985), Thwaites and Oakey (1985), Mackintosh (1986), Nijkamp (1986), Aydalot and Keeble (1988) study the impacts of high technology on regional development, while Giaoutzi and Nijkamp (1988), Hepworth (1989) and Brotchie *et al.* (1990) present material on the increasing importance of information processing and telecommunications.

Two reports by the Commission of the European Communities are indispensable reading for anyone interested in regional development in Europe: *The Regions in the 1990's* (1991a) addresses the still large disparities in economic development among the regions of the Community. *Europe 2000: Outlook for the Development of the Community's Territory* (1991b) is the authoritative policy document for the changing responsibility and objectives of transnational regional policy in Europe.

Another group of books deals with problems of specific types of regions. The volume by Rodwin and Sazanami (1991), *Industrial Change and Regional Economic Transformation*, deals with problems of the restructuring of old industrial regions in western Europe. Robinsons's (1990) *Conflict and Change in the Countryside* looks into problems of rural regions in industrialized countries *vis-a-vis* fundamental changes in agricultural production and increasing environmental pressures.

CHAPTER 8

URBAN AND RURAL FORM

Definition

The process of spatial restructuring in Europe is not restricted to the relative growth or decline of whole regions, but also affects the internal structure of regions, the relationship between cities and their hinterlands and the spatial organization of daily life within urban and rural regions. Nowhere has the impact of changing transport and communications technologies been felt more strongly than in this field. In some sense it can be argued that the form of our present cities and villages is largely a product of transport technology, in particular of the automobile, but the reverse is also true: that current settlement patterns dictate the still growing demand for mobility and transport infrastructure in urban and rural areas.

The links between this chapter and the chapters on goods transport (Chapter 9), passenger transport (Chapter 10) and communications (Chapter 11) are obvious. However, the form of cities and villages is equally the expression of long-term developments in population (Chapter 3), life-styles (Chapter 4) and the organization of production and distribution (Chapter 5). In a more general sense, urban and rural form is the mediating variable between these broader socio-economic trends and transport and communications.

With these considerations in mind, this chapter reviews trends in urbanization and urban growth and decline in Europe, patterns of urban deconcentration and of recent inner-city revival but also over-agglomeration, as well as emerging problems of small and medium-sized and peripheral towns and rural regions with special reference to the two-way interaction between urban and rural land-use patterns and transport and communications.

Trends

After the decline of the Mediterranean urban system, the history of the European city started again in the tenth century. Agricultural surplus production enabled the division of labour in the form of crafts and trade which flourished in larger more specialized settlements. However, the real start of urbanization was linked to the coincidence of rural over-population and labour demand in the emerging urban industries in the late eighteenth and early nineteenth centuries. Since then the proportion of people living in cities in Europe has grown by a factor of ten and is still growing. Yet today there are still large differences in urbanization in Europe between, for example, 63 per cent of the population living in cities over 50,000 in Great Britain and less than 20 per cent in Portugal (see Figure 8.1).

The 1950s and 1960s were a period of rapid urban growth all over Europe. High birth rates, continued rural-to-urban migration due to the mechanization of agriculture and a first wave of international labour migration from southern to northern Europe resulted in massive urban expansion, mostly in the form of new satellite towns or large high-rise suburbs.

In the 1970s the urbanization rate in several countries started to decline. After two hundred years of continuous growth, cities were facing a decline in population and later also of employment, first in their inner cores, later also in their suburbs and eventually in the whole urban area. At the same time small and medium-size cities at the less urbanized fringe started to grow. In some European countries there was even a virtual reversal of the urbanization process with peripheral areas increasing their share of total population and employment. Counter or deurbanization tendencies can be found primarily in the most urbanized countries in north west and central Europe, whereas in the Mediterranean (and in eastern Europe) the urbanization process fuelled by rural-to-urban migration prevails (see Figure 8.2).

However, not all cities in a country follow a common pattern. The new urban hierarchy divides cities across national boundaries into 'successful' and 'unsuccessful' cities. Successful cities are the few 'world' cities and the limited number of 'European' financial and cultural centres which flourish with the intensification of international trade and information flows, some regional centres with a favourable combination of locational factors and the large number of small and medium-size towns at the fringe of large agglomerations. All other cities which have none of the features necessary to compete in the increasingly international market for new economic activities are the losers, with old industrial cities and remote rural towns suffering most from urban decline.

The common experience for both winner and loser cities has been spatial

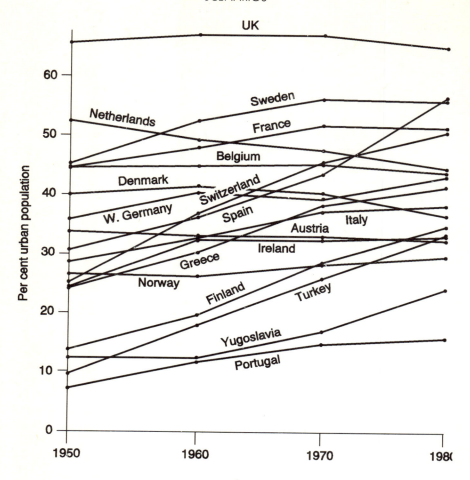

Source: OECD, 1986c

Figure 8.1 Proportion of population in cities over 50,000. The real start of urbanization was linked to the coincidence of rural overpopulation and labour demand in the emerging urban industries in the late eighteenth and early nineteenth centuries. Since then the proportion of people living in cities in Europe has grown by a factor of ten and is still growing. Yet today there are still large differences in urbanization in Europe between, for example, 63 per cent of the population living in cities over 50,000 in Great Britain and less than 20 per cent in Portugal.

deconcentration (see Figure 8.3). The evolution of transport systems made possible the expansion of cities over a wider and wider area. In particular the diffusion of the private automobile after the war brought low-density suburban living within the reach of not only the rich. However, suburbanization was not caused by the car, but is a consequence of the

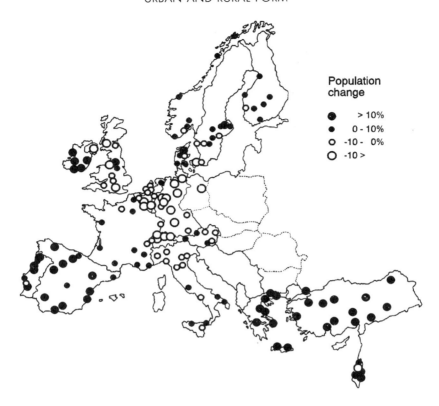

Figure 8.2 Urban growth and urban decline in Europe, 1970–85. In the 1970s the urbanization rate in several countries started to decline. At the same time small and medium-size
cities at the still less urbanized fringe started to grow. In some European countries there was
even a virtual reversal of the urbanization process. Counter or de-urbanization tendencies
can be found primarily in the most urbanized countries in north-west and central Europe,
whereas in the Mediterranean (and in eastern Europe) the urbanization process fuelled by
rural-to-urban migration prevails.

same changes in socio-economic context and life-styles that were also
responsible for the growth in car ownership: increase in income, more
working women, smaller households, more leisure time and a consequential change in housing preferences (see Figure 8.4). Yet the car certainly
contributed to pushing people out of the city centres through congestion
and lack of parking space and noise and pollution, as did housing short-
ages and high land prices. Offices and light industry and retailing started
to decentralize later, following either their employees or their markets or
both, taking advantage of attractive suburban locations with good transport accessibility, ample parking, and lower land prices. In particular,

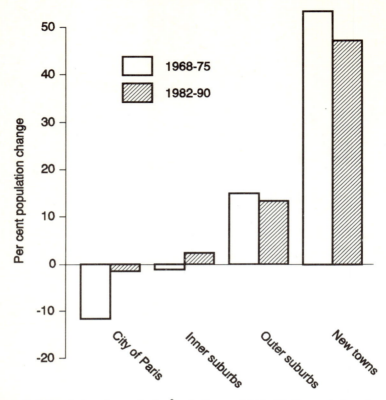

Figure 8.3 Population change in the Île de France, 1968–90. The evolution of transport systems made the expansion of cities over a wider and wider area possible. In particular the diffusion of the private car brought low-density suburban living into the reach of not only the rich. Suburbanization was not caused by the car, but is a consequence of the same changes in socio-economic context and life-styles that were also responsible for the growth in car ownership: an increase in income, more working women, smaller households, more leisure time and a consequential change in housing preferences.

greenfield shopping centres have become a threat to inner-area shopping, and 'airport cities' draw service activities from the traditional centre.

The results of the deconcentration process are generally considered to be negative: longer work and shopping trips, more energy consumption, pollution and accidents, problems of public transport provision in low-density areas and excessive land consumption (see Figure 8.4). Moreover, it made access to car travel a prerequisite for taking advantage of employment and service opportunities, thus contributing to social segmentation. All over Europe, cities have undertaken efforts to revitalize their inner cities through restoration programmes, pedestrianization schemes and new public transport systems. In some cases these efforts have been remarkably successful. Recent figures suggest that the exodus from the

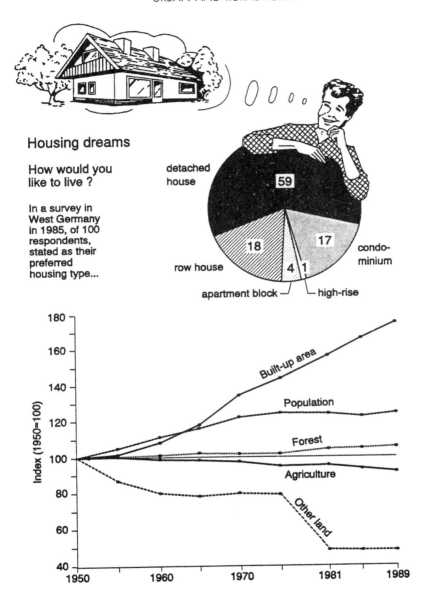

Housing dreams

How would you like to live ?

In a survey in West Germany in 1985, of 100 respondents, stated as their preferred housing type...

detached house — 59

row house — 18

apartment block — 4

high-rise — 1

condominium — 17

Index (1950=100)

Built-up area

Population

Forest

Agriculture

Other land

180
160
140
120
100
80
60
40

1950 1960 1970 1981 1989

Figure 8.4 Housing dreams and suburbanization. In a survey undertaken in 1985 in West Germany, 77 per cent of all respondents said they would like to live in their own house (less than 40 per cent actually do). This is one of the reasons why despite little growth in population the built-up area in Germany has almost doubled since the war. However, growing land consumption is not only caused by dispersed housing; all urban activities tend to consume more and more land per unit of activity.

inner city may have passed its peak and that there may be a 'reurbanization' phase.

The renaissance of inner-city living demonstrates the vitality of the European city with its history and architectural heritage as well as the increasing diversity of urban life-styles in Europe. Never before have the urban centres of Europe attracted so many visitors to their historical monuments, museums and theatres. Never before have there been so many new glittering shopping arcades, boutiques and restaurants. However, under present market conditions, reurbanization invariably involves a replacement of the original low-income inner-city population by more affluent residents who are able to afford the increased rents ('gentrification') thus adding to social segregation within the city. At the top level of the urban hierarchy, in London and Paris, but also in cities like Brussels, Frankfurt, Munich and Milan, this has, in conjunction with a new boom in the demand for luxury hotels, office space and convention facilities fuelled by massive real estate speculation, led to exorbitant increases in real estate prices and building rents, which threaten to make the central areas of these cities unaffordable as places to live for the majority of the population.

At the other end of the urban-rural spectrum, small rural towns and villages in agricultural regions are threatened by one of three developments. In the wider hinterlands of the agglomerations, in resort areas or in the mountains they lose their rural character as a result of the outer wave of urban sprawl, most notably in the form of second homes for people living in cities. In areas of high-intensity agricultural production, they suffer from what may be called 'industrialization without urbanization', with agrofactories dispersed over a monofunctional plain, severe pollution problems and a distinct lack of amenities. Finally, in remote regions at the European periphery, their very existence is threatened by depopulation and the collapse of the agricultural economy altogether, with irreparable ecological consequences.

All the above trends are still in effect and none of them has completed its cycle. In the trend scenario, therefore, a continuation of current tendencies seems to be most likely. In essence, this means a widening of the gap between winner and loser cities, further decentralization of activities within urban areas and an erosion of rural settlement patterns. Without strong government intervention, the latter trend is likely to be reinforced by intensified competition in the Single European Market.

Scenarios

Which of these trends will be dominant and which will have the strongest impacts on the future of urban and rural form in Europe? Which of the

A *Growth scenario*

The race for the world's highest office building was finally decided by the completion of the 330-storey *Murdoch Tower* on Canary Wharf in 2012. The cluster of shimmering skyscrapers in the Docklands had surpassed the old City, although land in the Docklands was more expensive than at prime locations in Shinjuku. The typical major city today consists of three distinct parts: the *international* city centre where incomes, rents, shops and restaurants follow standards set in Tokyo, London and New York; the *middle-class* suburbs where the majority of wage earners live a comfortable life; and the *underclass* ghettos where jobless people and illegal immigrants are supported by welfare. In smaller cities, the middle-class uses the redecorated town centres to demonstrate its wealth. Manufacturing has moved out to industrial parks near motorway exits (see *Goods transport*). The countryside close to agglomerations and in resort areas has practically disappeared.

B *Equity scenario*

The urban policy of the European government was based on the conviction that a variety of urban, suburban and rural life-styles would best achieve the overall goal of equitable opportunity. Therefore the government pursued a policy mix of incentives and regulations to reduce agglomeration diseconomies at the top of the urban hierarchy and to give financial aid where the market process failed to provide momentum. Through property gains taxes and a value capture legislation it was possible to keep land speculation in the large cities within limits. The richer countries of Europe joined funds to support the restoration of inner cities in countries which could not afford to do so themselves, making sure that the housing so recreated was reserved for the original tenants. A similar programme was set up to redirect some of the revenues from agricultural import charges into the restoration of decaying rural towns in southern and eastern Europe (see *Population*).

C *Environment scenario*

The ecological revolution has nowhere left its mark so strongly as on urban life-styles and urban form. In 1997 the European Parliament stated the resource-conserving city as the main goal of European urban policy. Car ownership was heavily taxed, making car use extremely expensive (see *Environment*), so most people looked for a home close to their work. Retail and services, and eventually also manufacturing, decentralized in order to be nearer to their customers and workers. Dispersed, low-density suburban settlements integrating living and working appeared everywhere; proposals for high-rise developments were turned down. Public transport was made more efficient; the resulting reduction in car traffic made it possible to downgrade some trunk roads or convert them into parks.

resulting scenarios will be most or least desirable? The three seed scenarios offered to the respondents (see Scenario Box 8.1) represented three extremes.

The *growth scenario* extrapolates developments observed today in the most advanced world cities such as London, New York or Tokyo. The central areas of these cities increasingly serve a few high-profit activities at the expense of other urban functions. Densities in the centre are excessive and land prices exorbitant. The *Murdoch Tower*, London Docklands and Shinjuku (a fast growing office district of Tokyo) are metaphors for this monofunctional development. Yet the scenario also points to the dark side of the shining world city in the form of social segregation, disadvantage and inequality. Smaller cities seem to do well if they are connected to agglomerations, but more peripheral ones do not. Manufacturing has ceased to be ugly, noisy and polluting; the 'industrial park' near motorway exits symbolizes the brave new world of high-tech and logistics. There is a brief reference to the urban sprawl which has devastated the countryside around agglomerations and in resort areas. One can surmise that more peripheral rural regions will suffer from depopulation and neglect.

The *equity scenario* is based on the expectation that a future European government will follow a strategy of decentralization and regional equalization. To implement this strategy a broad spectrum of anti-agglomeration policies are applied to redirect economic activity from the core urban areas to cities and regions at the European periphery. Within the core agglomerations, land speculation is effectively controlled and with it the social fragmentation resulting from high land prices. The scenario is built on a system of redistributional transfers: the richer countries help the poorer ones to preserve their urban heritage, and small towns in the rural European periphery receive compensation to enable them to survive in the harsher competitive climate of the Single European Market. In its optimism that the principle of cooperation and solidarity may coexist with the pre-eminent paradigm of competition and growth in a future Europe, this scenario may appear naïve. But it remains appealing as a significant alternative to the almost exclusively growth-oriented policy paradigm of the present national governments in Europe and the European Community.

The *environment scenario* illustrates how city and countryside might look in 2020 if resource conservation is the main goal of settlement planning. All policies are designed to halt the increasing spatial division of labour and bring work, living and shopping, education and recreation closer together again and so reduce the need for intraurban and intraregional travel. The scenario foresees dispersed, low-density, relatively self-contained suburban settlements rather than a return to the high-density integration of activities found in medieval and nineteenth-century cities. Consequently, retailing and services and employment in general also need to disperse. One might call this scenario the dissolution of the traditional city. Both city and countryside converge to a new 'rurban' settlement continuum. The scenario rightly points out that to make this work requires a high penalty on travel, especially car travel, and more efficient forms of dispersed public transport.

The views of the experts

Are these the three basic options for urban and rural development in Europe? In general the experts agreed that the three scenarios broadly cover the range of possible future developments. However, there were a number of comments and suggestions to elaborate or modify the proposed scenarios. The most significant of these addressed the issues of centralization/decentralization, dispersal and mobility and the future of inner cities.

Centralization/decentralization

There was agreement that the question of the 'optimal' urban concentration and density are crucial for both the equitable and the resource-conserving future city. However, there seem to be two contradictory views of the dominant forces at work. On the one hand, it was concluded that urban growth tends to increase both density at the centre and the spatial division of labour through further dispersal and that this leads to further growth of mobility of goods and passengers. On the other hand, there is the potential of new telecommunication technologies to substitute physical transport and of new high-speed transport systems to open up 'new spaces for growth' in distant areas such as eastern Europe and the Mediterranean 'Sunbelt' thereby reducing the need for spatial polarization.

Another comment was that there is no uniform pattern of urban development all over Europe as different spatial trends coexist in different parts. So it is possible to speak of a growing specialization and polarization at the highest level of the urban heirarchy in Europe, but also of a diminishing distinction between city and countryside towards a 'rurban' continuum. Small and medium-sized cities, especially if they are well connected to the large agglomerations, may play a key role in absorbing the spill-over population from the urban cores and offer better living conditions than both large city and countryside. Yet just as there are 'successful' and 'unsuccessful' cities, so there will be winners and losers among the rural areas. On the one hand there will be a move to the villages displacing unwanted farmland, but truly peripheral rural regions will suffer from depopulation and poverty. To overcome the growing spatial disparities, some reconcentration of the power to plan and allocate resources and compensations at the national or regional level may be required.

Dispersal and mobility

It was noted that advances in transport and communications have loosened the locational constraints acting on social or economic activities and that this tends to lead to a greater spatial division of labour and more dispersed settlement patterns. The dilemma is that dispersed settlement structures consume more space and make efficient public transport difficult. And even if homes and work places are closely integrated, will people take advantage of it? More likely, it was argued, they will continue to commute over long distances. Teleworking may make some commuting redundant, but so far the social and psychological disadvantages of working at home have prevented this becoming common practice. Teleshopping might reduce the number of shopping trips; however, the opposite seems to be the case as leisure shopping in city centres as well as in outlying shopping malls is becoming more and more popular.

Clearly a better integration of urban land use and transport planning is in order, but in the short term policies directly addressing transport behaviour were felt to be more effective. Here basically the same policies were suggested as in passenger transport (Chapter 10). Most frequently such policies restrict car use to its true social cost through petrol taxes or road pricing in sensitive areas. In addition, the subsidization of public transport, the support for bicycles, new forms of paratransit and electric vehicles and a more even spread of peak traffic through flexible working hours were suggested. However, in the long term, land-use policies to control urban sprawl and preserve the existing green spaces must be applied. One way of doing this would be to promote small and medium-sized towns to create a polycentric settlement system.

Inner cities

It was argued that it is as yet unclear whether inner-city decline is the dominant pattern or whether there is a renaissance of urban centres. If anything, 'reurbanization' is a qualitative phenomenon. It usually consists of the replacement of a traditional activity by a more profitable one, for instance of housing by offices, or of low-income housing by up-market apartments ('gentrification'). Needless to say, this kind of displacement has to be avoided in the equity scenario. That, however, may be impossible under present land market conditions. The experts agreed that land speculation has developed into one of the primary obstacles for rational urban land use and transport and planning, and that only radical constraints such as property-based capital gains taxes and priority purchasing rights for local government will be able to control the devastating effects of an exploding land market.

Which scenario?

With these reservations, about half of the respondents (31) felt that the growth scenario was the most likely to occur given existing trends, followed by the equity scenario (19). However, only a small minority (6) also chose the growth scenario as their preferred future option. Remarkably, more respondents opted for the equity scenario (27) than for the environment scenario (19). This may indicate that the environment scenario in this field was felt to be too radical compared with the equity scenario, or that environmental concerns have to stand back if basic issues of equality have not been resolved. However, some experts felt that the three scenarios do not need to be in conflict; that it should be possible to enjoy the benefits of urban growth and still safeguard equal living conditions and protect the environment. One respondent argued that the growth scenario will eventually change into the environment scenario, but that first 'some very nasty things' must happen before the general public will support environmental measures. 'Today it is just a dream'.

Speculation

The extent to which the three scenarios are useful descriptions of fundamental options for urban and rural development is demonstrated by an experiment conducted by the National Physical Planning Agency of The Netherlands for the *Randstad*, the huge metropolitan area between Amsterdam and Rotterdam (see Figure 8.5). In the exercise three basic alternatives for the spatial development of the Randstad were sketched out, which were associated with the terms 'freedom', 'equity' and 'quality' (TNO, 1990). The exercise shows that distinct spatial organizations and policy combinations could be designed that corresponded to the objectives expressed by the three terms. The 'freedom' scenario most closely resembles the growth scenario; it is characterized by polarization between tertiarized core cities and dispersed residential settlements and an increased spatial division of labour and mobility. The 'equity' scenario preserves the 'green heart' of the Randstad at the expense of higher densities in the rim cities with high-speed traffic corridors between them. The 'quality' scenario, like the environment scenario, builds on a small-scale integration of working and living in self-contained distributed settlements in close contact with nature. The authors expressed concern that the first scenario was the one most likely to occur and clearly favoured the last one.

In the exercise reported here the experts also feared that the growth scenario, in the absence of intervention, would be likely to come true, and the vast majority of them were critical of this development. However, neither of the two alternative scenarios seemed to be desirable enough to attract a clear majority of votes. There was more agreement about what one expects to occur and dislikes than about what one wishes to see.

This result only reflects the immense diversity of problems and issues affecting cities in Europe today. Cities differ not only in size but also in history and location, function and economic specialization and the extent of their growth or decline. London and Paris and the other large cities in the 'Blue Banana' have different problems from the cities at the European periphery. So have declining old industrial or port cities and the fast growing cities in the south. Each type has its own particular set of problems and concerns and cannot completely be compared with others.

Yet there are trends that will affect all cities in Europe in one or another way. The ongoing internationalization and globalization of economic processes will reinforce the competition between regions and cities in Europe, and cities that do not find their specialized niche in the wider European market will decline. High-speed transport and telecommunications networks will create new hierarchies of accessibility and information, and cities not connected to them will be disadvantaged. Within metropolitan areas, local governments or even suburban centres will continue to compete with the central city for infrastructure and private sector investment, and those that do not will fall back. Finally, it is likely that large uncontrolled immigration from peripheral European regions and outside the European Community will aggravate the competition for housing and income opportunities in cities.

In other words, the future of cities and regions in Europe will be determined by *competition*. In order to survive, cities will be forced to grow at the expense of other cities and their hinterlands, because in a competitive market the only alternative to growth is decline. Unharnessed urban growth, however, increases the disparity between cities in the European core and at the European periphery and between individual parts of the metropolitan region, is wasteful in terms of energy and other resources, and leads to congestion, exaggerated land prices, social inequalities and negative environmental impacts in the cores of the agglomerations.

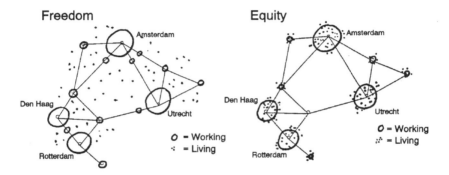

Source: TNO, 1990

Figure 8.5 Alternative scenarios for the Randstad, Holland. The National Planning Agency of the Netherlands sketched out three basic alternatives for the spatial development of the Randstad, the metropolitan area between Amsterdam and Rotterdam (TNO, 1991). The 'freedom' scenario is characterized by polarization between tertiarized core cities and dispersed residential settlements and increased spatial division of labour and mobility. The 'equity' scenario preserves the 'green heart' of the Randstad at the expense of higher densities in the rim cities and with high-speed traffic corridors between them. The 'quality' scenario builds on small-scale integration of working (werken) and living in self-contained distributed settlements in close contact with nature.

Under these circumstances, policy-makers responsible for the development of urban and rural form in the countries of Europe will have to choose between two courses:

either   to continue to promote the 'winner' cities by concentrating high-level infrastructure, public facilities and subsidies in the largest cities, by rewarding wasteful and environmentally harmful modes of transport and by not intervening in the destructive competition of large and small cities against each other and in the exploitation of natural resources through urban sprawl through a *laissez-faire* regional planning policy;

or   to engage in an active anti-agglomeration strategy by promoting peripheral small and medium-sized cities through modern transport and communications infrastructure, by decentralizing government agencies and by promoting local industries through seed funds and tax incentives, while at the same time constraining further urban sprawl at the outskirts of large agglomerations through greenbelt policies or other development controls.

Cities, in particular those in the richest countries of Europe, will have to make up their minds whether the pursuit of ever increasing material wealth can be their ultimate goal, or whether through cooperation with other cities and a better distribution of benefits and burdens among their own citizens a higher level of well-being for everybody can be achieved without never-ending quantitative growth. Under this perspective urban policy-making would include policies for the equalization of housing and education opportunities, for the spatial reintegration of work and home locations and for the promotion of environmentally less harmful transport modes. Villages and rural regions, on the other hand, in particular those at the periphery of large agglomerations, need to consider the long-term consequences of a sell-out of their open space to quick-profit urban developers and investors.

Is there a chance for urban and rural development that is both equitable and in balance with nature? The answer is not easy. To give up the principle of competition altogether would not be a real option as much of our wealth and much of the excitement of large cities are based on competition in the market economy. However, it would be a worthwhile goal to explore solutions where controlled growth can be guided in a socially balanced and environmentally sustainable way. Technology such as cleaner cars, more efficient public transport and less polluting manufacturing techniques might help. Better land-use planning to achieve a more rational organization of urban space may also help, but only in the long term. New or improved forms of cooperation between cities or between cities and their region may overcome the zero-sum game of competition. The multicultural, polycentric, affluent and high-tech and yet ecologically sustainable urban-rural region (see Text Box 8.1) is still an open option.

EXCURSION TO TOMORROWLAND

Thomas Özgal slowly drifted across Schalke in a north-easterly direction without leaving his flight altitude of 2,500 m. His destination was the 'Green Hand', a recreation park in Dortmund's north. It was Friday, the 24th of July, 2039. Underneath him was the greenbelt, where once had been the industrial heart of the Ruhr area. Large parts of the Emscher river basin between Duisburg, Dortmund and Bergkamen, that in the past had been the core of the Ruhrgebiet, had been converted into woodlands and parks. Several decades, many billions of Euro-Dollars and substantial amounts of imagination had been necessary to overcome the long-term impacts of industrialization.

Behind a bank of clouds in the north-east he saw the first suburbs of Datteln, Herten and Oer-Erkenschwick, which belonged to the problem areas of the otherwise so successful Eurometropolis Rhein-Ruhr. There lived the low-tech workers who worked mainly on Fridays, Saturdays and Sundays in poorly-paid part-time jobs which could not be replaced by robots or unpaid self-help. There lived the foreigners lured to Germany by the government's immigration programme and the second-hand dealers and migrant workers. Thomas Özgal's grandfather who was born in Erzurum in Turkey had lived in such a suburb. He himself, however, had not been there since his childhood, and there was nothing that could attract him.

Thomas Özgal was now approaching the solar airfield at the fringe of the 'Green Hand'. In front of him was the lawn runway with the large fly-and-bike parking area. He softly touched down and taxied his plane to the parking position assigned by the tower. Traffic on this morning was heavy. Thomas Özgal took his rent-a-bike and headed for work. He was part-time director of Senior GmbH, a small firm specializing in collecting information for elderly people and redistributing them via cable television to interested clients. Their information network included 9,000 small and medium-sized companies in the Eurometropolis Rhein-Ruhr.

The meeting had ended surprisingly early, so Thomas Özgal could phone his wife to plan for the evening. Karin Özgal was in charge of training Turkish transport engineers at the Ruhrtransport AG. She only had to go down to the hall to view the events of the day and make a booking over a terminal. Karin Özgal chose a performance in the nuclear museum of Hamm-Uentrop about the long-term impacts of Chernobyl and the wrap-in of the high-temperature reactor by

Christo, which in 2000 had heralded the end of the nuclear age in the Federal Republic. The Özgals decided to take the eco-taxi to Hamm, though their Piper Solar was equipped for night flights, because after the performance they wanted to have dinner in Unna, which during the 1990s had developed into a centre of Italian culture, design and trade.

Text Box 8.1 Extracts from *Excursion to Tomorrowland* (Kunzman, 1989). Is there a chance for urban and rural development that is both equitable and in balance with nature? The multicultural, polycentric, affluent and high-tech and yet ecologically sustainable urban-rural region is still an open option. Kunzmann's scenario for the Ruhr area in Germany *Excursion to Tomorrowland* (1989) shows how a formerly grey smoke-stack industrial area might be transformed by the year 2039.

Further reading

The history of cities has been the topic of many great writers. Here it is appropriate to start with Mumford's *The City in History* (1961); after more than thirty years this is still a fascinating *tour d'horizon* of the origins and transformations of urban culture. Equally worth reading today is Jacob's (1962) *The Death and Life of Great American Cities* because it already, thirty years ago, pointed to problems that still today plague our cities. A more contemporary example of historical writing on cities is Hall's (1988) *Cities of Tomorrow*, which is a splendid account on the development of urban planning thought in this century.

In contrast to these historical approaches to grasp the evolution of cities there are the products of scientific efforts to disentangle the interdependencies of factors and influences determining the internal spatial organization of urban regions. Of these, the contributions of the urban economists and the urban sociologists stand out. Urban economists see the city predominantly as a system of highly interdependent markets such as the land market, the housing market and the regional labour market. In these markets economic agents behave following economic principles just as in any other market. It is impossible to give even an impression of the large volume of research conducted in the field of urban economics. However, the paperback by Vickerman (1984) may serve as a primer and guide to this literature.

Urban sociologists, and some geographers, see the city primarily as the scene of social interaction and conflict. Again, as with urban economics, it is impossible to do justice here to the multitude and diversity of traditions and approaches developed in urban sociology and related disciplines. However, this work is of particular interest here because, in contrast to the urban economics literature, it has shown an intense interest in questions of equity and the distribution of advantages and disadvantages in cities. Two classical works in this field are Castells' (1977) *The Urban Question* and Harvey's (1973) *Social Justice and the City*. Hirschfield (1991) presents recent empirical evidence of growing poverty and social and spatial marginalization in western cities. The contributions in Smith and Williams (1986) analyse displacement processes associated with gentrification. Badcock (1984) looks into the specific kind of inequity built into cities where inner-city residents own fewer cars but suffer most from traffic noise and pollution.

Quite another category of writing on cities deals with the transformations currently going on in cities in Europe and elsewhere. Van den Berg et al. (1982) is an influential empirical study of urbanization and suburbanization processes, that is the *internal* concentration and deconcentration of population, in European cities. This book put forward the theory of urban development phases referred to in this chapter and has substantially influenced the discussion on the future of cities. Champion (1989) is a more recent collection of papers on counter-urbanization. Hall's *The World Cities* (1984) and his two books on London, *London 2000* (1963) and *London 2001* (1989) are outstanding examples of city case studies.

More recent work has, like the literature on regional development, focused on the impacts of modern technologies on cities. Two good collections of papers on the impacts of technological change on urban form are Brotchie et al. (1985) and (1987) and Castells (1989), with special reference to the role of information technology on cities. Kunzmann and Wegener (1991) summarize the impact of these developments on the urban system of Europe.

CHAPTER 9

GOODS TRANSPORT

Definition

Goods transport means the movement of all forms of raw materials, building materials, energy, food and industry products for distribution. Goods transport is a derived activity which reflects the interplay of the forces described in preceding chapters: modern industrial economies rely on efficient and reliable door-to-door transport links between mines, farms, industries and markets (see Chapter 5), and the organization of the flow of goods profoundly influences regional development (Chapter 7) and urban and rural form (Chapter 8). Goods transport is one of the foundations of consumption and life-styles in affluent societies (see Chapter 4). However, with growing volumes of goods transported and the increasing shift to goods transport by road, goods transport in Europe today also becomes a major environmental issue (see Chapter 6). Ships, cargo trains, trucks and aircraft are the major modes of goods transport. Energy can be transported by pipeline and electrical grid. Modern goods transport relies heavily on telecommunications (see Chapter 11).

Trends

The general trend over the past centuries of goods transport in Europe has been growth. This growth has paid for the development of ever larger and faster vehicles and their infrastructure. Life in modern industrial societies strongly depends on efficient goods transport. The variety of fresh food in stores and on breakfast tables is the result of efficient logistic chains. The materials and components in household products have been transported between many industries before they are distributed as market products.

Sea transport along the coasts and on inland waterways was the dominant mode of goods transport in Europe until two centuries ago. Sea transport still has its competitive edge in bulk cargo. Supertankers transport crude oil from the Arabian Gulf to west European refineries, modern diesel barges are used on European rivers and canals. Roll-on/roll-off (RoRo) ferries connect the Mediterranean islands, the British Isles and Scandinavia with the European continent. The contribution of inland waterway traffic to total goods transport varies between countries depending on their endowment with navigable rivers and canals (see Figure 9.1). In the future the completion of the Rhone-Rhine, Rhine-Main-Danube and Oder-Elbe-Danube canals will improve the market share of water transport.

Europe has a well developed rail system for goods transport. Railways are faster than coastal shipping and barge transport and thus made inland locations of industries possible and decreased the importance of port cities. Without railways the rapid industrialization of Europe in the second half of the nineteenth century would not have been possible.

Today railway goods transport is losing ground against the competition of the truck. On average railways attract less than 20 per cent of all goods transport and much less in countries such as Great Britain, Italy and The Netherlands (see Figure 9.1). National railway companies are making great efforts to halt the erosion of their market share by streamlining services, direct overnight freight connections and aggressive freight rates or by promoting various forms of combined road and rail transport such as 'piggyback' trains (where the whole truck is entrained) or swap bodies, containers or trailers on rail cars. However, against the unsurpassed speed and flexibility of door-to-door truck service none of these policies have yet been really successful.

Trucks and vans on roads are at present the dominant mode of goods transport in Europe. In Britain, the homeland of the railway, the lorry has almost entirely replaced goods transport by rail and the same is true for Italy. Since 1970 truck transport in western Europe has more than doubled, while rail and water transport have stagnated or even declined (see Figure 9.2). There have been several reasons for the rise of goods transport by road. The road network represents a fine mesh of accessibility to all corners of the continent, and even forests and farms are connected via gravel roads to the main road network and thus to industries and population centres. Every year the road network has expanded in line with economic growth, in particular the motorway system (see Figure 9.3), while at the same time rail networks in all countries have been constantly reduced. Moreover, with low diesel prices and taxes and roads being financed by the government, road goods transport has never covered its true social and environmental costs, in contrast to the railway which has to finance its own infrastructure.

129

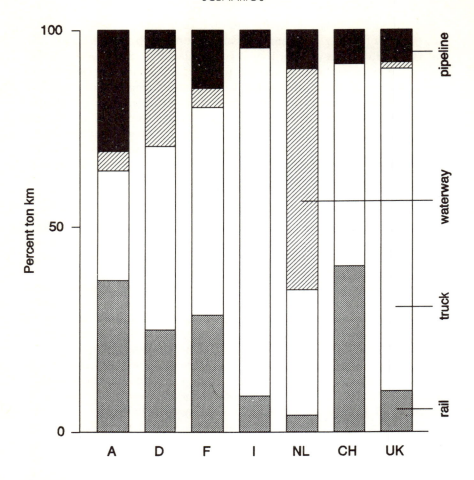

Figure 9.1 Goods transport by mode in seven European countries, 1986. There are marked variations between countries in the use of different modes for goods transport which reflect their geographical and historical circumstances. For example, waterway transport is heavily used in Germany and The Netherlands but not elsewhere. Similarly rail transport is an important mode in Austria and Switzerland but little use is made of rail in Britain and Italy. In both the latter countries over 80 per cent of goods are transported by road.

The success of the truck is mainly due to its advantage in door-to-door speed, flexibility and reliability. Modern market economies with their multitude of interdependencies and a dispersed settlement structure could not exist without efficient, unbroken, door-to-door transport links. Logistic systems linking procurement, production and distribution processes substitute warehousing functions by 'just-in-time' delivery of small shipment sizes and hence often rely on flexible small and medium-sized vehicles.

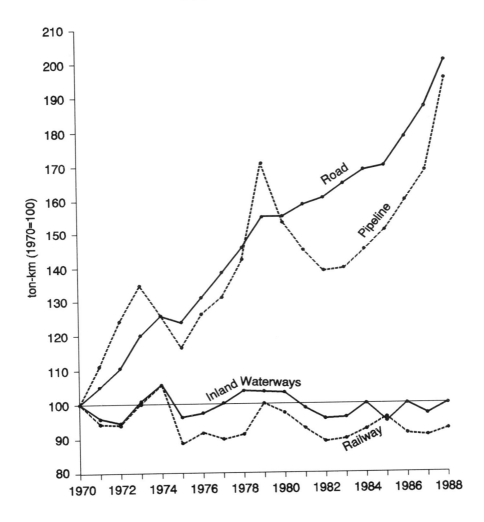

Source: European Conference of Ministers of Transport, 1989a

Figure 9.2 Goods transport by mode in western Europe, 1970–88. Since 1970 truck transport has more than doubled while rail and water transport have stagnated or even declined. Today truck transport accounts for nearly three-quarters of the ton-km transported.

Increasing value per ton makes it more and more important to have a driver with personal responsibility for the transport process. Even long-distance sea and rail transport would not be possible without regional road transport at their start and end. As costs for warehousing and reloading increase, long-distance trucking also increases.

Figure 9.3 Motorways and European highways. The road network represents a fine mesh of accessibility to all corners of the continent and makes even peripheral regions easily accessible by truck. The map highlights the concentration of motorways in the highly industrialized regions of north-western Europe.

The term 'logistics' implies the integration and optimization of procurement, production and distribution processes through information and telecommunication technology subject to objectives such as cost, flexibility and reliability. An unbroken sequence of logistic processes is called a logistic chain (see Figure 9.4). Logistics have not only fundamentally changed the production and distribution system but also the transport

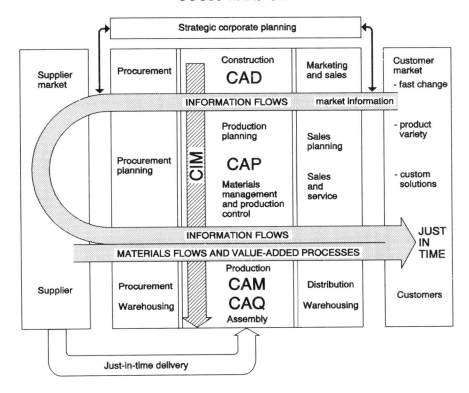

Source: Läpple, 1990

Figure 9.4 Logistics in manufacturing. The 'logistic revolution' has fundamentally changed the production and distribution of goods. The term logistics implies the integration and optimization of procurement, production and distribution processes through information and telecommunication technology subject to objectives such as cost, flexibility and reliability. An unbroken sequence of logistic processes is called a logistic chain. The figure highlights the integration of computerized subsystems such as computer assisted design (CAD), planning (CAP), manufacturing (CAM) and quality control (CAQ) by computer-integrated manufacturing (CIM).

industry itself. Integrated logistic freight and fleet management includes three dimensions: the actors involved, the infrastructure and the process-oriented view, goods and information flows (see Figure 9.5).

The overall trends in goods transport can be described as follows. Both the volume of cargo and the value of goods transported increase as the standard of living rises and as products are becoming more refined and complex. Slower transport modes are gradually being replaced by faster ones. Advances in information processing and telecommunications give rise to new possibilities for complex logistics systems including multimode

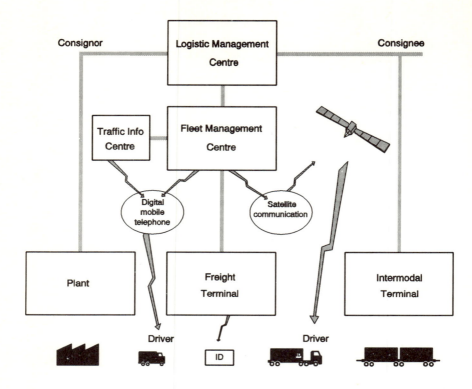

Figure 9.5 Integrated logistic freight and fleet management scenario. This scenario makes use of road transport informatics currently being developed. Satellite communications are available today and a European digital mobile telephone network is planned. Freight identification numbers are transmitted from vehicles to roadside beacons so that their progress can be continuously monitored at the logistic management centre.

transport. The logistics revolution within industries is spreading out to inter-industry transport.

Railway goods transport is facing further decline unless measures are undertaken to alter its competitive disadvantage. Client-specific customized services, further progress in standardization and containerization, new freight centres and computerized freight information systems may improve the situation of railways on the freight market. Many see the introduction of high-speed rail as a possible turning-point in the downward trend of rail freight transport. The Single European Market will increase average transport distances and reduce customs procedures and thus may offer new opportunities for rail or combined road-rail transport.

In particular, transit countries like Switzerland and Austria are considering new combined transalpine links such as the 'moving motorway' through the Gotthard and Lötschberg base tunnels.

The lorry will also benefit from the new freedom. It has been estimated that a further rise of road goods transport of 30 per cent will be due to the Single European Market and this will probably present a major congestion problem on motorways, in particular in transit countries such as Switzerland and Austria. Nevertheless there will be a relative decline in road infrastructure investments in the highly industrialized countries in central Europe. Countries at the European periphery still have a strong interest in new road infrastructure, but will increasingly face environmentalist opposition.

Air transport is the fastest, youngest and fastest-growing goods transport mode. It has an advantage for long-distance transport of high-value industrial products as well as for courier service parcel delivery. Goods transport by air is primarily 'underbelly' transport in jet liners as a fringe operation to passenger transport. However, increasingly pure freight jet planes are also in operation. Air freight will continue its upward trend with the growing internationalization of the European economies. Deregulation of air transport will lead to greater competition and reduced freight rates.

Scenarios

The general trend in goods transport seems to be a persistent growth in volume, speed and distance, a continuing shift from slower modes of transport (rail) to faster ones (lorry and air) and an increasing dependency of production and distribution processes on highly computerized just-in-time door-to-door logistics.

In the face of these trends, one can imagine several scenarios of the development of goods transport in Europe. The *growth*, *equity* and *environment* scenarios below describe three possible corridors of future development (see Scenario Box 9.1). Which of them will be most likely and which will be most desirable?

The *growth scenario* projects a continued increase in goods transport fuelled by a vigorous growth of the European economy. This spurs transport industry to invest in more and larger vehicles in order to distribute the wealth. Trucking is the fastest growing mode. This is expected to lead to congestion due to a lack of infrastructure, but also to private investments in toll motorway networks.

A   *Growth scenario*

The vigorous growth of the European economy, the introduction of the Single European Market and the deregulation of the transport industry have given a boost to truck and air freight operations at the expense of rail and sea. During the 1990s, the dramatic increase in road freight traffic caused congestion on motorways and unacceptable levels of noise and pollution in cities. After the turn of the century, however, a gigantic building programme for an entirely new network of European toll motorways, bridges and tunnels financed by private capital improved transport capacity to meet demand. Most truck freight is handled via strategically located freight centres by a small number of transnational carriers. Today railway freight service has practically disappeared. Air cargo is increasingly by special jet and propjet freight aircraft operating in transport networks from former military airports to avoid air congestion over cities and to allow night operations.

B   *Equity Scenario*

An equity-minded transport policy faces a dilemma: freight transport by truck is the more egalitarian as it provides almost equal accessibility by road all over Europe even to small communities. On the other hand, trucks are environmentally harmful, wasteful in terms of energy and take cargo away from the railways. The result of this dilemma has been a *laissez-faire* European transport policy. Truck traffic continued to increase and the deficits of the railways became staggering. However, the European government neither consented to a privatization of the railways (which would have killed railway freight services anyway), nor did it invest in new motorways or licensed private companies to build toll roads, with the result that by the end of the century road traffic in and near agglomerations nearly collapsed.

C   *Environment scenario*

The 'green' solution has been to restrict heavy road traffic and to increase its cost by mandatory use of catalysts and cleaner and more refined fuels such as LPG, compressed natural gas and methanol. Freight operations by air have been limited by regulation and heavy taxes. Large-scale industrial activities with their high dependence on goods transport are gradually replaced by more local production. Thus rail has kept a larger share of goods transport, with the present capacity of the rail infrastructure as the upper limit. Recently new goods aircraft propelled by liquid hydrogen and clean fuel engines for road vehicles have become available. These quiet and clean 'ecological vehicles' are used to support a more decentralized society. A dispersed network of low-density settlements is gradually replacing some of the environmentally ineffective 'city deserts' with their pollution.

The *equity scenario* is accompanied by problems with overutilized and congested roads, railways with deficits and congestion at harbour and airport terminals due to low infrastructure investment. In an equity perspective transport is seen as a public utility which should be provided for everyone and everywhere. This dilemma leads to a *laissez-faire* attitude towards goods transport. The growth of goods transport is not accompanied by adequate infrastructure investment.

The *environment scenario* restricts heavy road traffic and promotes rail transport. Those transport modes that are most polluting, air and road, are most constrained by taxing and regulation with the effect that more freight is transported by rail—even though this involves extra costs and delays through reloading because almost all goods transport starts and ends by road. In the very long perspective however, the 'green' solution includes hydrogen-fuelled trucks and aircraft, making these modes more environmentally friendly.

The views of the experts

Are these the three main options for Europe? Most respondents accepted the scenarios as feasible paths of development under different assumptions. Their comments, however, indicate the complexity of the goods transport issue. There are no simple growth or equity solutions. According to one respondent, 'The equity scenarios are not very likely. In the end some form of polarization will take place. In the growth scenario for goods transport there ought to be room for expanding or building special rail-cargo lines in addition to the truck and aircraft networks'.

The modifications suggested by respondents relate chiefly to four main lines of development: to impose environmental constraints, to revitalize the railways, to revitalize market forces and to improve logistics.

Impose environmental constraints

Given the discussion in Chapter 6 it should come as no surprise that there was general agreement that environmental constraints should be imposed on goods transport. However, there were considerable differences in opinion as to the likely impact of these constraints. Some respondents argued that an environmental collapse can be prevented by the adoption of combined rail/road transport and logistics. But even if this occurs, governments will be forced to adopt more environmentally friendly policies. The anti-pollution measures taken in California and Japan are good examples.

Some respondents felt that present pollution and congestion problems can be alleviated through clean technology and road pricing, charging for the use of sensitive road space like in city centres. One expert argued that the trucking industry does not pay for its secondary and higher-order costs for infrastructure and the environment. In a deregulated market social costs would have to be included. If long-haul trucking has to pay higher charges for infrastructure, congestion and emissions, then reloading and multimodal transport become more competitive.

It was also argued that the trend scenario of lorry transport continuing to grow at the expense of rail seems almost unavoidable given its economic logic and the complex spatial structures which lorry transport makes possible. However this view does not take account of the 'green' pressures on the goods transport sector. 'These pressures are more likely to be accommodated by further technological improvements (e.g. controlling engine emissions) than by forced adjustments in the mode of goods transport or in spatial structure'. Nevertheless, the overall conclusion from these comments is that the imposition of environmental constraints will have a profound effect on the growth of goods transport over the next thirty years.

Revitalize the railways

Many respondents were more optimistic about the future of rail freight transport than was implied in the scenarios. Several argued that by the privatization of railway lines they can be made competitive. Others claimed: 'Stop pretending that railways need protection!', 'Invest more in railway modernization. The privatization of railways will not kill freight services', 'Furthermore, rising fuel prices will result in more railway goods transport.'

It was also argued that strategic rail investment is likely to have large secondary effects on modal choice for goods transport. 'Although no mention is made of freight transport through the Channel Tunnel', the tunnel represents a rail link between the British Isles and the continent. When it is in operation, there will be an additional incentive to use rail transport. Similar effects can be hoped for as a result of the north-south tunnelling through Switzerland that is planned to counteract the environmental problems from heavy lorries. When these tunnels are operational, goods can be transported by rail from the United Kingdom and northern Europe to southern Italy all the way by train.

Revitalize market forces

Deregulation and the introduction of pricing mechanisms were suggested by many respondents as a means of revitalizing market forces. However, it is important that social costs are included in market prices. In the opinion of one commentator, 'market forces can be used to avoid congestion and concentration because these two adverse effects are costly'. Consequently 'the taxation principle for transport should be marginal social costs'. Other forms of usage taxation and investment were suggested. For example, 'petrol should be taxed heavily to accelerate clean fuel substitutes—and provide "electronic freeways" for heavy vehicles'. In all these cases the cost incentive is used to change traffic behaviour as well as raise the capital for investment. Under these circumstances it can be argued that 'expanding rail cargo lines, in parallel to road and air transport, can be a viable solution for Europe 2020'. 'Air cargo will be mainly important as a transcontinental mode and thus not within Europe'.

Improve logistics

An interesting vision of logistics in the future city was provided by one respondent:

> The European goods transport system in the year 2020 will be dualistic. In large conurbations goods traffic will be banned and replaced by a comprehensive automatic underground goods handling system. At the outskirts of these conurbations, there will be terminals for the interchange of goods with the European high-speed goods network (European Inter-urban Goods Transport System), a TGV-like system exclusively for goods transport. Distribution within the cities, from the public stops of the underground, is done with the help of battery-driven cars.

At the European level this vision could take the following form.

Include logistic centres in European infrastructure planning. Develop and subsidize the introduction of multi modal transport, between a few locations (fifteen to forty in all Europe), that are faster, more reliable but not necessarily cheaper in the beginning than trucking. Implement a multimodal open booking and tracking system for goods like AMADEUS for air passengers. Internalize environmental costs in the user charges according to congestion and equity between regions. Hence environment and equity consequences are internalized in the market processes for developing new technologies, products and services as well as the use of them. Regulations can be reduced to a minimum.

Other alternatives suggested by respondents include combined freight-passenger trains, a vacuum tunnel solution as a new transport mode to decrease the surface transport of goods, 300 km/h goods trains and fast (130 km/h) catamarans, in competition with fully automated truck freeways.

Several respondents pointed out that a European policy on infrastructure development is needed to achieve these objectives. One respondent pointed out that *The Integrated Road Transport Environment* (IRTE), envisaged by European R&D programmes on the use of modern road transport informatics should be implemented. 'New developments in road transport informatics and telecommunications have reduced paperwork and improved the efficiency of freight transport through better fleet management and better coordination of modes through multi modal transport etc. Therefore a more balanced picture of freight transport has been obtained which in addition to cleaner engine technology has improved efficiency while keeping environmental impacts to a minimum'.

Which scenario?

The largest group of the sixty respondents (26) felt that the growth scenario was the most likely. The equity scenario ranked as second (19) and the environment scenario was seen as the least probable (8). As the most preferred scenario, however, the environment scenario had strong support from almost two-thirds of the respondents (37). The pure growth scenario was preferred by only a minority (10). This outcome indicates a development from the established growth paradigm to a more 'green' solution being more justified in the long run. Some respondents (6) thought that growth and environment need not necessarily be in conflict and opted for a combination of the growth and environment scenarios. Surprisingly, very few respondents (4) preferred the equity scenario.

Speculation

In summary, the economic forces behind a more deregulated freight transport market will favour each mode where it has comparative advantages. However, without dramatic changes in the structure of freight rates, this is unlikely to strengthen multi modal transport using rail or inland waterways for part of the distance. If, on the other hand, environmental issues continue to grow as a dominant value for societal development, a pricing mechanism taking account of the social and environmental costs of the different transport modes such as road pricing will become necessary. A disfavoured mode, like heavy diesel trucks of today's technical standard, will have to pay a road usage fee reflecting their damage to the road, their noise and their CO, SO_2, NO_x and particle emissions. Such a pricing mechanism would be a strong incentive to the fuel and engine industries to develop clean propulsion alternatives.

If the growth scenario is the most probable, it must at least be accompanied by concerted action to decrease pollution. Furthermore, as lorry transport will remain the dominant mode of goods transport, the trucking industry must be the main target for limits to growth due to environmental constraints. In a future road transport environment, with all cars being equipped with catalytic converters, diesel trucks will stand out as even more noisy, smelly and air polluting than today. To counteract this, cleaner fuels have to be adopted for lorries, especially in urban areas. Delivery vans may have to switch to catalyst engines using reformed diesel fuels, liquid petroleum gas (LPG) or natural gas blends, or combinations of them.

To revitalize the railways seems at first glance a clean alternative to trucking. However, one has to bear in mind that the railway network is not finely meshed enough to reach all locations. Most goods transported by rail begin and end their journey on lorries or delivery vans. Goods transport by rail will also in the future be economical only for long distances, and even here the capacity of the rail network will impose limits to large shifts of goods from road to rail. Furthermore, the electricity used for the 'clean' rail transport is dominantly generated by power plants using coal or heavy fuel oils. This means that, though rail transport is much cleaner than road transport, it is not the ultimately clean solution. Nevertheless, rail transport has the great advantage that every tonne carried by rail relieves the growing congestion on Europe's motorways.

New telecommunications and road transport informatics functions can help to improve fleet and freight management. Today a very large proportion of lorries and railway cars run empty. A deregulated goods transport market will allow cabotage, i.e. the right to take goods between countries other than the origin and destination of the prime transport. With appropriate computerized planning tools, it will be possible to utilize better existing fleet capacity, take advantage of favourable mode combinations, optimize the routing of transports in order to avoid congestion and minimize delays. All this will contribute to a better utilization of transport infrastructure.

However, all these policies will find their limits if the volume and distance of goods transport in Europe continues to grow explosively as in the recent past. It will therefore be necessary not only to make goods transport more efficient and less environmentally damaging, but also to look into the deeper causes of this incessant growth. Here questions of industrial organization, internationalization and just-in-time logistics and their long-term impacts need to be addressed, and it may well be that policies in the field of regional development (see Chapter 7) and urban and rural form (see Chapter 8) directed at reducing the need for moving goods around may contribute more to minimizing the negative impacts of goods transport than policies addressing transport issues directly.

So for the policy-makers responsible for planning the future goods transport system in Europe, the policy options will lie between two extremes:

either   to continue to reward road freight transport through low taxation of lorries, aggressive motorway construction, neglect of rail freight services outside large agglomerations and retarded introduction of stricter emission controls for trucks, and to promote further the haulage of products that can be produced and consumed regionally such as vegetables, flowers or beverages, and to tolerate or even subsidize the establishment of dispersed, space-consuming production facilities relying on transport-intensive just-in-time logistics;

or   to reduce road freight transport through fair taxation of lorries taking account of the damage of lorries to roads and the environmental costs following the 'polluter-pays' principle, to introduce immediately rigorous environmental standards for lorries and traffic restrictions for lorries in environmentally sensitive areas and congested areas in cities during rush hours or at night in residential districts, to invest substantially in combined road/rail transport facilities and services and to reduce the volume of freight by the relocation of heavy process industries, disincentives for excessive just-in-time logistics schemes and the promotion of regional, short-distance distribution networks.

Neither of these two extremes will probably be implemented in its pure form. The first alternative would soon become unbearable through its unsolved congestion and pollution problems. The second alternative would require the willingness of people to sacrifice a small part of their affluence and convenience for a sustainable environment. This awareness exists to varying degrees in all European countries but many national governments, and in particular decision-makers at the European Commission, still seem to place the highest priority on the global competitiveness of the European industry and hence on its transport needs. However, at the same time the number of policy-makers who support a new balance between transport modes is growing. To revitalize market forces to establish this new balance seems to be the politically most feasible path of development. The important thing is that the pricing mechanism is not left to regulate itself but is shaped by public policy decisions in accordance with the values of access, mobility and a clean environment. A regulated market mechanism would have to be flexible enough to adapt to local and national policies as well as to different values on the quality of life.

Further reading

The growing importance of goods transport is reflected by the increasing number of books dealing with issues relating to goods transport in Europe. One of the most interesting publications in this field is the recent (1990) report produced by the Group Transport 2000 Plus set up by Karel van Miert, Transport Commissioner of the Commission of the European Communities, *Transport in a Fast Changing Europe*. The report is a broad *tour d'horizon* of all aspects of transport in Europe including goods transport and it well reflects the variety of perspectives and interests associated with this field.

Other publications are more specialized. Button's (1982) *Transport Economics* and Faulks' (1990) *Principles of Transport* are two textbooks covering the fundamentals of transport economics and transport planning. Barde and Pearce (1990) and Quinet (1990) present the state of the art of assessment of external effects and social costs of transport. Nijkamp, Reichman and Wegener's (1990) *Euromobile* gives a comparative view of issues in transport research in different European countries.

McKinnon (1989) and Cooper et al., (1991) describe the 'logistics revolution' and the changes it brings to goods transport in the wake of changes in production technology and the location of manufacturing (see Scott and Storper, 1986). One of the two companion volumes of this book, *Transport, Communications and Spatial Organization* (Giannopoulos et al., forthcoming) contains several contributions on the same topic, including a chapter by Wandel and Ruijgrok on freight transport and spatial organization and production. The second volume, *Logistics Platforms and City Logistics* (Ruijgrok et al., forthcoming) is a comprehensive treatment of multimodal logistics and their spatial implications. The growing importance of telecommunications in goods transport is the subject of a report by the European Conference of Ministers of Transport (1989b).

With the Single European Market approaching, policy questions of goods transport are attracting increasing attention. Button and Gillingwater (1986) and several papers in Nijkamp and Reichman (1987) as well as the new book by Button and Banister (1991) deal with issues of deregulation of transport operations and the financing of transport infrastructure. Whitelegg (1988) is a compilation of useful information on transport policy in the countries of the European Community. An industrial perspective on transport policy can be found in the following three documents published by the European Round Table of Industrialists in Brussels: *Keeping Europe Mobile* (1988), *Logistics and Transport in Europe* (1990) and *Missing Networks* (1991). As might be expected, these publications highlight the need of rapid improvement of European transport infrastructure networks and a coordinated European transport policy.

CHAPTER 10

PASSENGER TRANSPORT

Definition

Passenger transport is a form of transport with high demands on availability, just-in-time arrival and comfort. Personal mobility has become a central value of modern society, and the freedom to move with ease between home, work, service and leisure has become a characteristic of a modern life-style.

Like goods transport (see Chapter 9) passenger transport is a derived activity. It is the result of all the factors and trends discussed in Chapters 3–8. The size of the population (see Chapter 3) gives a first indication of the demand for passenger transport in a society. The prevailing life-styles (see Chapter 4) define personal mobility behaviour, and regional development (see Chapter 7) and urban and rural form (see Chapter 8) define the pattern of trips. With rapid growth in personal mobility, passenger transport has become a major environmental problem (see Chapter 6).

Passenger transport can be classified by purpose (work, shopping, business, education, leisure, etc.) or by mode. This chapter reviews the trends in passenger transport with respect to its three dominant modes: road, rail and air transport.

Trends

The transformation of an agricultural society to a modern industrial society has been followed by an enormous growth in personal mobility. The average daily travel distance per person has increased in the order of one hundred times during the last century. During the last decades the passenger car has become the dominant travel mode as measured by passenger kilometres travelled. This has been accompanied by suburban sprawl around cities, making inhabitants dependent on the automobile for all

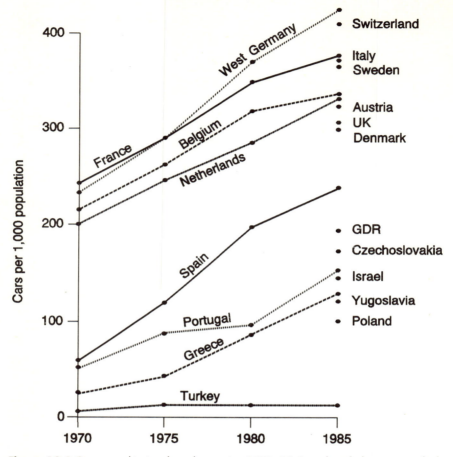

Figure 10.1 Car ownership in selected countries, 1970–85. Even though there are marked variations between countries, car ownership has increased markedly since 1970. The highest levels of car ownership can be found in countries such as France, West Germany, Switzerland and Sweden. Lower levels of car ownership exist mainly in southern and east European countries such as Greece and Poland, Turkey and Yugoslavia.

travel but for trips to the city core. The relative importance of travel by bus and rail has decreased as the car has claimed supremacy for local and regional travel.

The dependency on the automobile was illustrated during the 1973–4 period of high oil prices. A doubling of fuel cost did not drastically stop automobility (see Figure 10.1). Growth was dampened, but in essence the car had already become an established and necessary part of the life-style for most families. Today 80 per cent of all passenger-km are made by car (see Figure 10.2). The success of the car is due to the freedom of movement and almost universal usefulness it offers. It is equally good for short

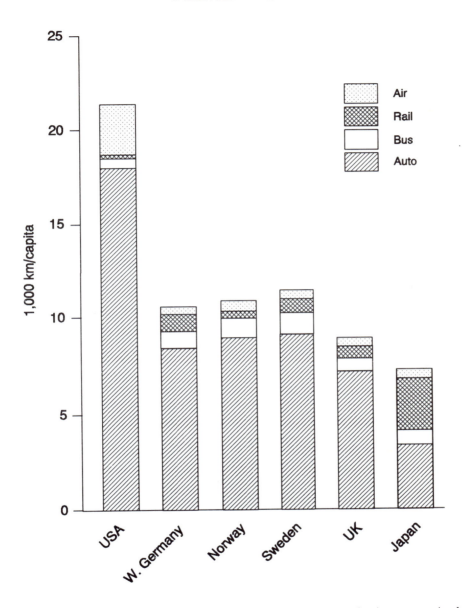

Figure 10.2 Passenger travel by mode in selected countries, 1986. The dominant mode of passenger transport in most countries apart from Japan is the automobile. Since 1973 the number of kilometres per capita travelled by car has risen sharply at the expense of other modes. In most European countries personal mobility, measured in terms of kilometres per capita travelled each year, is still less than half that of the United States.

as for long distances, can be used for carrying people and goods, requires no changes of mode and only minimal planning before starting a journey. Consequently, all forecasts have notoriously underestimated the growth of car ownership and even in countries with the highest levels of car ownership no saturation is in sight, though the highest growth rates are now in east European countries.

During the last two decades a new trend has been observed regarding car ownership. The large family car is gradually being replaced by smaller vehicles for individual travel. This relates both to the increased number of single-parent families and the increase in work out of homes for women. The case for collective travel between domestic areas to industrial sites is being lost in favour of a multitude of individual trips at different times of the day between different working places, service and education facilities, leisure activities and homes. In this process, work-related trips decrease in favour of shopping and recreation-related trips, among them tourist travel. The rise of the car has been accompanied by, and has partly caused, suburban sprawl around cities, making inhabitants depend on the automobile. Living in low-density suburbs at the urban periphery is only possible with a car. However, in densely populated areas the automobile has already shown its ultimate limits in the form of congestion and unacceptable levels of pollution and noise. In many countries it has become apparent that solutions to urban traffic problems can no longer consist of further expanding the road network but in a synergetic mixture of a variety of policies such as taxation, user charges, traffic constraints, pedestrianization, and promotion of public transport. It is beginning to be felt that the car and public transport should no longer be considered as competing but as complementary components of urban transport and that a part of the costs for public transport should be financed by the general public like the costs for road infrastructure.

Electric rail systems have the great advantage of being practically pollution free in operation. However, because of their high costs, subways and commuter rail lines have remained reserved for very large cities. For smaller and regional centres, light rail transit (LRT) systems offer affordable and efficient solutions with the same service quality as subways. In some countries LRT systems are operated by private companies. Buses serve as feeders to LRT or rail stations, but are the only mode of transport in small towns and rural areas. The greatest problem for public transport is the provision of adequate services in low-density suburbs or rural areas. Many countries are experimenting with new demand-responsive systems such as dial-a-bus or various forms of paratransit.

Some 10 per cent of all passenger-km produced can be classed as long-distance travel. For long-distance travel, train and airline services are the competing alternatives. With its dense railway network and generally short distances between its major cities, Europe is ideally suited for rail

New high-speed rail lines
Upgraded lines
Extensions and connections

0   300 km

Figure 10.3 The projected high-speed rail network in Europe. When implemented this network will provide an attractive alternative to both car and air for intercity transport in Europe. Sections of this network are already in service in France and Germany.

travel. Eurocity and Intercity trains offer efficient, comfortable and reliable, although slow, service between major European centres. An entirely new European network of high-speed trains on special tracks is envisaged for the 1990s (see Figure 10.3), although problems of compatibility between national systems still have to be resolved. This new level of rail infrastructure will change the geography of the continent by significantly reducing travel times between its major centres (see Figure 10.4). Already existing high-speed trains such as the French TGV successfully compete with domestic airlines. The German *Transrapid*, a MAGLEV system with linear motor, has still to find a domestic application. Besides investing in faster trains, many national railway companies are trying to attract more passengers by aggressive marketing, special discounts, service packages custom-tailored to individual groups and new combinations of services and trains such as 'Fly and Rail', 'Park and Rail', 'Rail and Drive', 'car trains' and 'hotel trains'. Another important trend is the development of railway stations and their surroundings as major centres of commercial activity, shopping and entertainment.

Airlines offer the most competitive travel for business and vacation trips over very long distances. During the last decade airline operations have experienced steady growth (see Figure 10.5). The growth in total world air travel has been some 6 per cent per annum during many years. Even the 'oil crisis' years did not change the basic long-term trend for growth. This indicates that air travel still is in its infancy, far from saturation levels. The projection is for a continued growth of 5 to 7 per cent per annum.

However, the future development of air travel in Europe will be largely determined by capacity restrictions on airports and air routes which already suffer from peak-hour congestion. There is a trend among airlines to offer more non-stop direct flights between destinations to avoid the congested airport hubs. Deregulation of passenger air travel will increase competition between carriers on major routes and probably lead to the creation of airline-specific hubs at individual airports to discourage inter-lining. Reductions in fares and a variety of special discount fares will stimulate the demand for air travel at the expense of rail, as will advanced computer reservation systems like *Galileo* or *Amadeus*.

Scenarios

Growth in personal mobility by car has been the dominant trend in Europe for the last five decades. This growth has been accompanied by suburban sprawl around cities. For medium-to-long distance travel, new high-speed rail systems compete with both the car and the airplane. Air travel is the fastest growing transport mode.

With these general trends in mind, three component scenarios have been formulated which reflect different life-styles and different demands on passenger transport (see Scenario Box 10.1).

The *growth scenario* embodies continued growth in personal mobility.

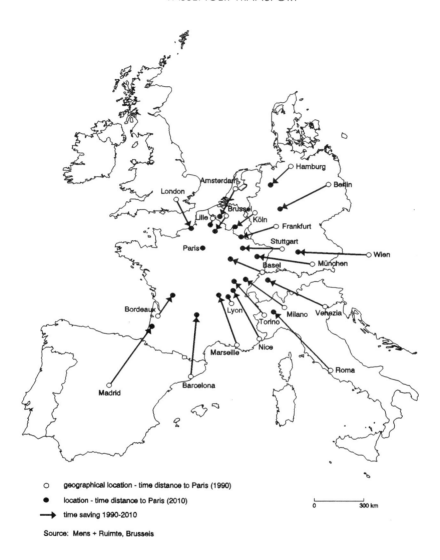

○   geographical location - time distance to Paris (1990)

●   location - time distance to Paris (2010)

→   time saving 1990-2010

Source: Mens + Ruimte, Brussels

Figure 10.4 Time savings by high-speed rail. The European high-speed rail network will change the geography of the continent by significantly reducing travel times between its major centres. The map shows by what extent destinations will 'come closer' to Paris once the full network of high-speed trains will be realized. For instance, travel times to distant destinations such as Berlin, Vienna or Barcelona will be almost halved. However, it should be noted that the assumptions about the time schedule of implementation underlying this map is a rather optimistic one.

151

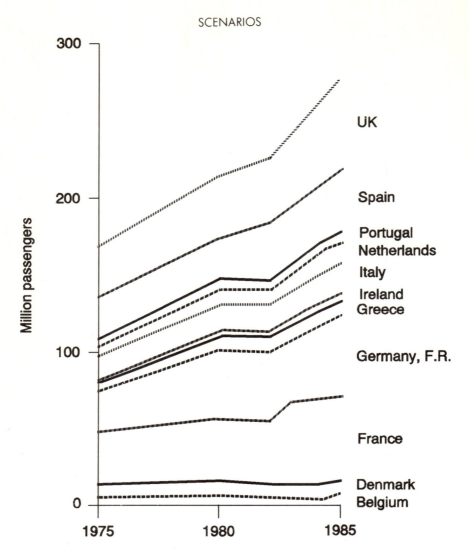

Source: Group Transport 2000 Plus, 1990

Figure 10.5 Growth of air passenger travel, 1975–85. All countries show a considerable increase in air passenger travel since 1970. In 1985 Britain led Europe in terms of the number of passengers carried by air, followed by France, West Germany and Spain.

All trends leading to more and longer trips, smaller households, more working women, more leisure time, increasing affluence and the growing spatial dispersion of activities are in effect. In the absence of strong regulatory policy, this leads to the supremacy of the car over marginalised public transport in urban regions and fierce competition between high-speed trains and aircraft in long-distance travel. Technological innovation serves

A *Growth scenario*

Never has there been such a mobile society. Fuelled by continued economic prosperity, individual automobility dominates local and regional travel. Road transport informatics and automated driving have made the use of the automobile feasible for a larger proportion of the population. Younger, older and handicapped people can now drive with the help of a semi-automatic road traffic machinery. Local public transport has been reduced to a feeder service of the automobile. Metros and buses between park and ride terminals and the city centre provide inner cities with work-force and consumers from the suburbs. High-speed trains link most European cities with each other. In long-distance high speed trains compete with airlines operating from more and smaller airfields with quiet STOL aircraft. However, the capacity of large airports and of the airspace over Europe has been reached.

B *Equity scenario*

As with goods transport, the government faced the dilemma of having to choose between the automobile, which brings accessibility to everywhere but at a heavy social cost, and the railway, which gives mobility to everybody but favours the agglomerations. Again the government has followed a mixed strategy by promoting both the car and the train, which did nothing to relieve road congestion or the cumulating deficits of public transport. More money is spent on regional rail service and rural roads at the expense of motorways and high-speed trains. Heavy user charges for cars entering the inner city are discussed as a possible means to discourage car use; however, the obvious equity problems associated with road pricing make this an unfeasible strategy. In air travel the government keeps tight control on traffic and fares yet allows cheap flights in reach of almost everybody.

C *Environment scenario*

The decision to ban the automobile was not popular, but the government convinced the public that the state of the environment made it necessary. The use of vehicles using petrol and diesel engines was reduced by heavy taxes on car ownership and petrol and rigorous emission standards. For the same reason jet travel was also restricted. A massive programme to revitalize public transport brought a renaissance of trolleybuses, trams, metros and electrical trains. Only recently some important breakthroughs in the field of new clean fuels have been made. Cars operated with hydrogen from solar collectors on homes will eventually provide environmentally harmless mobility for a dispersed society. However, this technology may take another century to mature.

to reinforce rather than counteract these trends by enabling even the young, old and handicapped to drive a car or by making even very short trips feasible for aircraft by short take-off and landing (STOL) technology. The scenario is tacit about the obvious consequences in terms of congestion or for the environment, except for a brief reference to limited air space capacity.

The *equity scenario* describes a dilemma familiar to many governments trying to reconcile equity and environmental objectives in transport policy. Promoting both the car and rail does neither contain the growth of the former nor solve the problems of the latter. Once automobility is considered a fundamental right of every citizen, policies to restrict it by monetary measures may be accused of discriminating parts of the population—an issue likely to become controversial in future debates about road pricing.

The *environment scenario* starts out as the radical green solution to transport problems: rigorous taxation, emission standards and restrictions on movement to constrain undesirable forms of travel and massive subsidies and promotion of desirable forms of public transport. No wonder that these experience a renaissance—the scenario suggests that not everybody may be happy. Yet there is also some confidence that benevolent technology will in the long run make environmentally 'harmless' mobility possible.

The views of the experts

Do these scenarios represent the options for passenger transport for Europe in 2020? In the opinion of the experts they covered most of the issues involved. However, it was also argued that passenger transport can develop in many ways and a wide range of modifications and improvements was suggested by the respondents. These can be grouped into three broad categories: the limits to personal transport, the potential of multi modal transport and the possibilities of changing mobility behaviour.

The limits to personal transport

There was general agreement among the respondents that congestion and pollution are the two key problems affecting the future of passenger transport. Congestion is spreading in Europe for both road and air transport (see also Chapter 6). Thus there is a limit to personal transport. Furthermore, the use of petroleum-fuelled engines causes pollution and depletes energy resources. Thus mobility must be limited by demand management methods and price mechanisms that help finance the development of clean fuels and energy conversion. As a respondent from France put it:

'Mobility is a key requirement for liberty, but the exponential development of mobility is not sustainable. So we have to combine mobility with energy and environmental constraints'.

Consequently it was argued that a growth of passenger transport can only be acceptable if the means of transport are inherently environmentally friendly. However, this still leaves us with the problem of city centres which lack the physical space for unrestrained personal mobility. The sheer size and number of cars is thus likely to define the ultimate limits to personal transport in city centres.

Not all the experts took such a pessimistic view. One British respondent argued that 'long-distance travel is likely to expand more slowly than many 1990 projections and local travel may swing away from cars to other modes'. Another felt that current pressures will force clean vehicles to come into operation before 2020. Others argued that there will be ultralight cars and bicycles in cities rather than more public transport.

The potential of multi modal transport

As in the case of goods transport (see Chapter 9) a greater use of multi modal passenger transport can substantially reduce pollution and congestion. However, as one respondent put it: 'multi modal total trip planning depends on efficient modal competition'. In other words, 'All passenger modes should be allowed to compete on equal footing' but at the same time, 'unbridled competition between modes should be replaced by a more cooperative/complementary stance and an increase in high-speed rail. This speaks for more demand-responsive forms of public transport'. Very small cars, besides high taxes, should be the means for financing better public transport.

With these considerations in mind, an alternative equity scenario was proposed by one respondent. This was labelled the 'impotence scenario'! 'For big cities at least congestion costs will force cities to invest in passenger transportation infrastructures, incorporating both private and public modes in combination as the only rational way of approaching the urban transport problem.'

There was also some support for the development of combinations of technological systems. Some commentators foresaw the breakthrough of private solar cars in 2005 and their massive production by 2020. Others argued that speed limits by technical means (75 km/h) will limit the car to urban agglomerations while public transport (rail and air) is provided as the interregional transport means. The proposal to 'introduce the magnetic levitated *Transrapid* rather than fast trains' highlights the potential of entirely new transport systems for long-distance travel.

Changing mobility behaviour

Some experts felt that external factors are likely to reduce the demand for passenger transport. They argued that the year 2000 may come to be regarded as a trend break in cultural values as new life-styles emerge (see Chapter 4). This could lead to drastic changes in attitudes. In the words of a German respondent: 'A new relationship to "time" must emerge—why has everything got to go so fast? Instead of promoting high-speed trains and air transport, the money should be used for "normal" public transport, both for short and long distances'.

Most respondents agreed that mobility is an important value in developed societies. Nevertheless, they felt that excess personal mobility should be restricted. Various techniques for demand management were suggested. For example, 'Apply road pricing to cars so that they are attractive only for long-distance travel and in regions with poor public transport'. Another respondent was in favour of 'heavy taxes on energy-inefficient cars to start with, leading to taxes on car ownership, transfer and use. Stimulate the extension of public transport networks'. There was also some support for the view that when you own a car, it is too inexpensive to use. A more balanced distribution of car usage might be achieved via user fees on personal mobility and the constructive reinvestment of those funds for more appropriate public transport.

It was also argued that there was an imbalance between transport modes. For example: 'It will be necessary to adopt adequate organizational forms in traffic on a European level (e.g. the use of certain airports for intercontinental traffic and other airports for traffic between distant European cities). It will be necessary to construct high-speed railway lines and rationalize passenger traffic in the big cities'. Statements such as these indicate the need to think more fundamentally about ways of reducing the demand for passenger transport than is currently the case.

Which scenario?

Nearly half of the respondents (27) saw the growth scenario as the most likely. The equity scenario got the support of 23 respondents. This implies that personal mobility is still regarded as a very strong value which is not yet satisfied. As the most preferred scenario, however, the environment scenario was supported by one-third of the sixty respondents (21). Only 13 respondents favoured the growth scenario and only a small minority the equity scenario. This result is an indication of a shift from the growth paradigm towards more environmentally conscious solutions. A substantial group of respondents (10) opted for a combination of the growth and environment scenarios suggesting that these two scenarios need not be in conflict in all cases.

Speculation

The development of passenger transport in the next thirty years will not follow the extremes of the seed scenarios. Many respondents observed that there is considerable overlap between the growth, equity and environment scenarios. The conclusion is: controlled growth under environmental constraints is a possible future provided that fuels and engines for road transport can be made cleaner and more environment-friendly. But even that solution leaves the congestion problem unsolved.

To solve both environmental and congestion problems requires not only cleaner cars but also alternatives to the car for sustainable personal mobility. No single transport mode can solve all problems, nor is there any single technological fix that can make transport altogether environment-friendly. Moreover, the time needed for changing the transport infrastructure is very long. Therefore it is necessary to use the existing transport systems as rationally as possible. Only a synergetic mix of policies can work. With the help of computerized traffic management systems, efficient pricing mechanisms and well-designed intermodal facilities, it will be possible to promote the use of public transport wherever feasible and combined transport modes such as park-and-ride for low-density areas where public transport cannot be provided at the necessary level of service. In inner-city and residential areas, and also in ecologically sensitive areas outside cities, the use of the private automobile will have to be even more constrained than many people today will consider acceptable. These measures will, of course, only be successful if they are accompanied by transport policies giving high priority to public transport investment and high-quality levels of service and land-use policies encouraging back-to-the-city moves and mixed land uses and a closer association of housing, work places and shopping and recreation facilities in urban areas.

For most people mobility is a very strong value. The environment issue has so far not gained the momentum to halt the long-term trend of growing mobility. It also would be opinionated to see mobility from a narrow environmentalist view only in negative terms. After all, in a dispersed society like ours, in many cases trips are necessary for personal interaction and cultural enrichment and cannot easily be dismissed as 'unnecessary' mobility. However, in the long run new attitudes and changing life-styles and values such as 'voluntary simplicity' and 'green life', in conjunction with land-use policies making long trips less necessary, may lead to a society with less travel. This development may be reinforced by the rapid diffusion of communication services which, for the first time in our civilization, will make it possible to combine rural living and working with full access to urban services and business opportunities. However, social habits of commuting or recreation do not change just because new opportunities exist. It can take a generation before the potential of telecommunication and computerization really changes the travel patterns in cities.

From the above speculations it can be concluded that the policy-makers responsible for planning and developing passenger transport have in principle two ways to go:

either   to promote further increases of car traffic through insufficient taxation of car ownership and car use, slow introduction of more rigorous environmental standards for cars and unconstrained continuation of road construction even in countries with fully developed road networks, while at the same time reducing investment and operating subsidies to railways and public transport; and to promote the trend to dispersed, space-consuming and transport-intensive settlement structures, which cannot be economically served by public transport, without concurrent decentralization of employment;

or   to discourage the use of cars by a differentiated system of taxation, user fees and road pricing taking full account of the environmental and social costs of car usage in general and in particular areas or times of day, while at the same time substantially improving the attractiveness of public transport by new investments, enhanced service and competitive fare structures (acknowledging the fact that public transport is a public good that needs to be subsidized); and to improve the spatial interaction of residences, work places, shopping areas and public facilities by promoting reurbanization and mixed land use.

The year 2020 is one generation away from today. How will our children want to lead their lives when at our age? In the survey particularly the younger people voted for the environment scenario. This may indicate that in the long run the environment scenario will win over the growth and equity scenarios. With the rapid changes in eastern Europe in mind, the year 2000 may become a major trend break. It could bring about major cultural shifts from the materialistic growth and equity paradigms towards the environment paradigm directed towards a new quality of life in accord with nature. For transport policy in Europe this means that a political strategy has to be adopted that in a credible way acts against the congestion and pollution problems of today. Otherwise the young people of today will be very disappointed with their governments by the year 2020.

Further reading

The historical context of the discussion on urban passenger traffic can be understood by reading the famous Buchanan Report (1963) *Traffic in Towns*. The problems today are not too different, but the solutions differ because today, hopefully, we know better. The special issue of *Built Environment* edited by David Banister (1989) with the topical title 'The Final Gridlock' illustrates the dilemmas of current urban transport planning and is a nice contrast reading to Buchanan. Mogridge's (1990) *Traffic in Towns: Jam Yesterday, Jam Today and Jam Tomorrow?* is a more provocative and slightly polemic commentary on the thirty years between Buchanan and Banister.

Many of the titles referenced here have already been mentioned in the previous chapter. Button's (1982) *Transport Economics* and Faulks' (1990) *Principles of Transport* cover as textbooks both goods and passenger transport, and also Barde and Pearce (1990) and Quinet (1990) are relevant here as sources on the assessment of external effects and social costs of transport. The reference to the *Euromobile* book by Nijkamp, Reichman and Wegener (1990) is also appropriate here and the discussion of transport policy and deregulation in transport in Button and Gillingwater (1986), Nijkamp and Reichman (1987) and Button and Banister (1991) applies to passenger transport as well.

However, one group of studies explicitly deals with personal mobility. Hägerstrand (1987) reinterprets his seminal contribution of the early 1970s, in which he established the vocabulary of action spaces and mobility constraints which has since become indispensable for mobility research. In Jansen et al. (1985) many of the issues dominating the debate about the limits to mobility were formulated.

Another group of publications looks at the interaction between transport and regional and urban development. Brotchie et al. (1987) contains chapters on the spatial impacts of transport technology including the final chapter 'The Transition to an Information Society'. Snickars (1987) and Salomon (1988) discuss the relationship between transport and telecommunications and the still unsolved question whether telecommunications tend to reduce or stimulate personal spatial mobility. Svidén (1983) argues that both mobility and telecommunications will grow in synergy and form a 'high-info-mobility-society'. All these issues are taken up again in Giannopoulos et al. (forthcoming).

There is a growing body of literature dealing with the changes going on in urban transport, the revitalization of public transport and the renaissance of the pedestrian in cities. Hall and Hass-Klau (1985) and Hass-Klau (1990) instructively summarize mostly continental experience, where these developments may have advanced somewhat faster than in Britain, for the British audience.

In long-distance travel, the rejuvenation of the railway in the form of the high-speed train is the focus of many publications. The report on rail network cooperation for the European Ministers of Transport (1989b) highlights the many difficulties still to be overcome before high-speed trains can freely change from one country to the other.

Finally, the recommendations for further reading on environmental aspects of transport given in the chapter on environment (Chapter 6) deserve to be repeated here: the various OECD reports (1986b, 1988a, 1988b), the country studies edited by Button and Barde (1990) and the growing literature on alternatives to the car such as Tolley (1990) or Vester (1990). Whitelegg and Holzapfel in *Transport for a Sustainable Future* (1991) summarize the discussion on the future role of the automobile and argue for a fundamental change in European transport policy.

CHAPTER 11

COMMUNICATIONS

Definition

Communication is the transport of information between human beings. Communication is the basis of social development. In a humanistic perspective communication with language, gestures and other non-verbal channels is a prerequisite for the individual to develop into a social being. Thus communication traditionally necessitates a meeting between the individuals involved. However, via modern telecommunications channels (telephone, radio, television) knowledge, social life-styles and cultural achievements can also be communicated to remote corners of society.

Telecommunications today pervade all fields of human life. The current transformation of life-styles (see Chapter 4) is co-determined by the ease of establishing and maintaining personal contacts over long distances. The modern economy (see Chapter 5) would not exist without efficient telecommunication links to suppliers and markets. Modern logistics in goods transport (see Chapter 9) and passenger traffic management and control (see Chapter 10) entirely depend on telecommunications. The most recent advances in telecommunication and information handling (facsimile transmission and computer networks) give rise to new levels of intensity and diversity in communication. This is sometimes described as the evolving information society.

Trends

Telecommunications are crucial to the government of a nation. The old Persian empire was governed via a horseback courier system. The ancient Greek society of port towns was held together through ships. The Roman empire utilized a road system linking the capital with its provinces. The British empire depended upon a safe sea mail service. Today's superpowers rely on satellite communication. History has many examples of governments controlling their people by restricting their communication with the outside world.

Communication is an essential component in the personal quality of life and for cultural development (see Chapter 4). There has also always been a strong interdependence between communications and economic development. Telegraph, telephone and radio were crucial to the development of railways, ocean shipping and aviation. These in turn boosted industrialization and international trade. In today's commercial life and industry telecommunications technologies such as telephone, telex, telefax and computer-to-computer data exchange are indispensable for the expansion of markets and the control of operations on an international scale (see Chapter 5). With these technologies the transport industry is able to manage sophisticated multimodal transportation systems (see Chapters 9–10).

Throughout this century communication technologies have developed rapidly. The microelectronics revolution of the last decades has created an explosion in telecommunication and information services (see Figure 11.1). Telecommunication services of all kinds are now within the reach of the individual for work and leisure due to the rapidly falling prices of microelectronics equipment.

The fast growth of the telecommunications market has been remarkable. In many European countries today practically every household has a telephone, although the differences between countries are still substantial (see Figure 11.2). Telephone communication started as a means to get voice contact with people. Today, a basic system with two copper wires connecting the subscribers to an exchange makes it possible to use a growing range of additional services like telefax and computer-to-computer data exchange (see Figure 11.3). The telephone network is being upgraded with broadband coaxial cables and fibre optics to cope with the increased traffic.

The newest and most impressive diffusion phenomenon is the explosive proliferation of facsimile machines (see Figure 11.3). Fax machines transmit an original from paper or a personal computer over the telephone network. Telefax is progressively taking over the functions not only of telex but also of ordinary mail for business communication such as orders, invoices and letters including drawings and maps and even pictures. The result is a considerable speed-up of response times in business activities—an increase in efficiency, but also in stress. Other new telecommunications technologies such as packet-switched networks or value-added circuits are spreading steadily but less visibly, others have experienced very uneven levels of acceptance in individual countries (e.g. videotext), while others have failed to pick up the market originally expected (e.g. video conferences).

Telecommunications are potentially a substitute for travel. Electronic signals are more easily moved over distances than physical goods or persons.

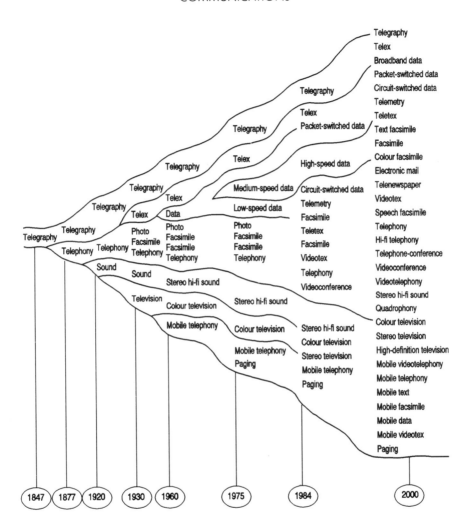

Figure 11.1 Telecommunications prospects for the year 2000. The microelectronics revolution of the last decades has created an explosion in telecommunication and information services. The diagram shows the evolution of 150 years of telecommunications from telegraph and telephone to the multitude of telecommunication options of the information society.

Figure 11.2 Telephone connections per 1,000 population in selected European countries, 1986. Norway and Sweden lead Europe, closely followed by Denmark and Switzerland. The lowest number of telephone connections can be found in southern European countries such as Spain, Portugal, Turkey and Yugoslavia.

164

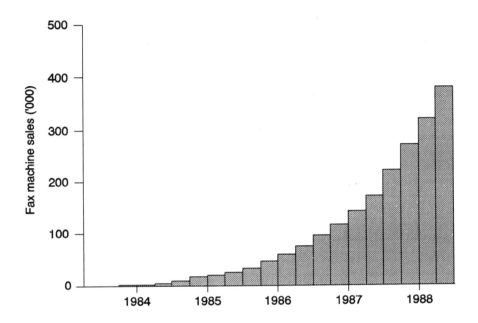

Figure 11.3 Sales of facsimile machines in the UK, 1984–88 (thousands). These figures for the United Kingdom indicate a spectacular growth in purchases of fax machines between 1984 and 1988. Purchases are still increasing in the United Kingdom and the overall trend mirrors that of many other European countries.

But more often telecommunications, rather than replacing the need for travel, act in the opposite direction by enabling a person or organization to be informed about new possibilities that require travel. For example, a television programme about a remote place can be the starting-point for planning a business or vacation trip. Telecommunications also make it possible for people to work at home with their computer terminal instead of commuting to work or do their shopping from home instead of driving to the shops. Direct personal contacts on business trips can be substituted by video conferencing or electronic mail. The substitution of travel by telecommunications may reduce environmental problems and energy consumption. Urban areas could also be more decentralized, and the disparities between central and peripheral areas might be reduced. However, the change from traditional business and working styles has been slower than many observers anticipated, and empirical evidence of these changes is still patchy and uncertain.

More fundamental and clearly observable are the impacts of telecommunications on goods transport (see Chapter 9). The first step was the optimization of truck fleet operations through radio-dispatch systems.

The next step involves implementing demand-responsive delivery systems or 'just-in-time' procurement systems to reduce the need for storage and warehousing. At the final level of integration even the production process itself becomes a link in the logistic chain. Telecommunications makes it possible for on-line sales information to be fed back along the logistic chain thus enabling production to respond almost immediately to changes in demand, replacing economies of scale by economies of scope (see Chapter 5). In terms of goods transport this means more trips, in particular more road trips by smaller vehicles.

Recent advances in information storage, retrieval and management in computers provided the starting-point for two breakthroughs in telecommunications satellite technology which enables telecommunications over global distances without cables, and fibre optics in conjunction with digital transmission and high-speed electronic switching equipment which opens the way for new commercially viable broadband communications networks.

The first application of the new fibre optics technology in Europe is the Integrated Services Digital Network (ISDN). The ISDN is a high capacity broadband communication system. It offers new services such as videotelephony and interactive television. ISDN will replace various telecommunications systems by transporting voice data, text and images at the same time. All the industrialized European countries are currently developing pilot versions of ISDN using conventional telephone lines working at a speed of 64K bits per second (narrowband ISDN). In many countries narrowband ISDN is already available in the major cities (see Figure 11.4). Narrowband ISDN is only capable of transmitting voice, data, text and slow-moving monochrome images; for fast-moving colour video transmission much higher band widths are necessary. In the next phase, transmission speeds of multiples of 64K bits per second using fibre optics are envisaged (broadband ISDN).

The introduction of broadband ISDN opens up the possibility of a variety of new information services beyond the ones in use today, such as videotelephony, two-way television, high-speed data communication, high-speed facsimile, colour facsimile, moving picture documents, teleshopping and high-resolution document and film retrieval. However, it is not certain whether the technological potential offered will generate a demand that is sufficiently large to justify the enormous investment that is required to install fibre optic networks and switching equipment for broadband ISDN. It is therefore expected that broadband ISDN will be available primarily for commercial users.

One question which is still largely unresolved is whether the introduction of new telecommunications networks will promote spatial concentration or deconcentration. It seems certain that because of the high investment necessary new services will be introduced first in countries and regions where

COMMUNICATIONS

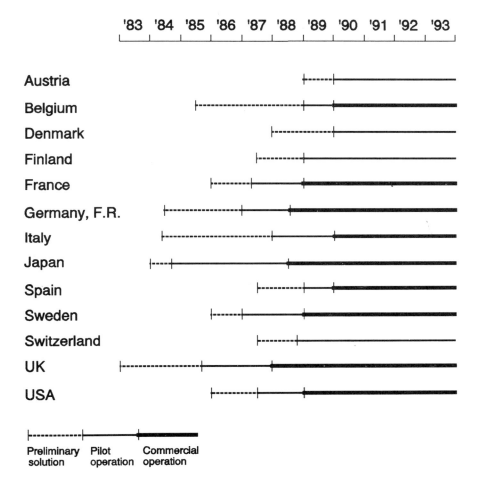

Figure 11.4 Plans for the introduction of the Integrated Services Digital Network in selected European countries. Most of the European countries have begun to implement the Integrated Services Digital Network. It is anticipated that the system will be in operation by the mid 1990s throughout most of Europe.

high-volume demand will make them quickly profitable. This will undoubtedly reinforce the already dominant position of these central regions. On the other hand new telecommunications services, once widely available, will reduce the locational disadvantages of remote regions by carrying information without delay between centre and periphery. This means that telecommunications could have a polarizing effect in the short run and an equalizing effect in the long run.

At the intraregional level the same argument holds, but it is anticipated

167

that the diffusion of the technology will be faster, so the equalizing effect is likely to be dominant. This means that outlying parts of metropolitan areas may become relatively more attractive compared to the core, and the metropolitan area could further decentralize. However, it remains to be seen whether the continued need for personal face-to-face contacts will outweigh this decentralization tendency.

Scenarios

Telecommunications technology has developed rapidly over the last decades and, together with pervasive computerization, has transformed nearly all fields of human life. It has speeded up response times in business activities and has the potential to displace partly slower modes of communication and even personal travel, although this has not yet materialized. Telecommunications have revolutionized goods transport and may have substantial impacts on the location and organization of industries and hence on regional development and urban form.

With these general trends in mind, the scenarios have been formulated to embody the *growth*, *equity* and *environment* paradigms (see Scenario Box 11.1).

The *growth scenario* predicts a continued growth in telecommunications capacity and new services in the first place for industry and business. One consequence of this is the more rapid implementation of Integrated Services Digital Network (ISDN) in city regions. This favours the central European countries at the expense of those on the periphery.

The *equity scenario* makes telecommunications charges independent of their distance from the information source as a matter of policy. This policy seeks to decrease the inequalities between life in the central and peripheral regions within each country, as well as in Europe as a whole. In this way, social contacts over long distances are eased, and the traditional culture of cities like theatre, films and exhibitions is distributed on an equal basis to people living in the rural areas and on the peripheries of Europe.

The *environment scenario* assumes that telecommunications are not in conflict with the environment. Instead, they can provide substitutes for travel and reduce the demand for paper products. Given good quality communication using ISDN, corporate networks can be established with employees working from home or neighbourhood office facilities. These dispersed living and working life-styles present a long-term alternative to those of congested urban centres.

A   *Growth scenario*

The unprecedented growth of the European economy is built on a vast volume of information transmitted over high-capacity fibre optics and satellite telecommunications networks. With Europe growing together and the economy becoming more and more globalized, each human activity is accompanied by a multitude of messages, enquiries, orders or confirmations sent through telecommunication networks. The 'wired' or 'information' city creates new life-styles and mobility patterns. It has become a status symbol of the information-rich, active and mobile to be informed, monitored, served and entertained by a host of 'intelligent' devices such as cellular telephones, pagers or voice-recognition hand-held computers. However, the telecommunication revolution has not been generated by private demand but by the needs of business. This has favoured the central European regions at the expense of the periphery.

B   *Equity scenario*

In the beginning it was feared that telecommunication infrastructure would mainly benefit large agglomerations. However, as the technology diffused to smaller settlements, telecommunication is now considered an important instrument to decrease differences between the industrial and governmental centres and the periphery. Via telephone, telefax and data lines firms even in remote villages have on-line access to industrial data bases. Villagers enjoy avant garde theatre transmitted from festivals in all parts of Europe as well as the Open University TV programmes via satellite. As an equity policy all telecommunications charges are independent of the distance from the information source—in contrast to former policies where long-distance calls were overcharged because the alternative was an expensive trip. This policy has contributed much to decreasing the inequalities between central and peripheral life.

C   *Environment scenario*

Telecommunication technology is not in conflict with the environment. It can be used to replace the growing consumption of wood for communication on paper and to substitute travel. Therefore the European government has supported the evolution of high-capacity communication networks and even promoted its experimental use for new forms of dispersed living and working in low-density but high-tech settlements. 'Electronic cottage' type villages in Kent, Brittany and Tuscany have become fashionable if not for data typists then for executives, writers, scientists and stockbrokers who are tired of city life but unwilling to give up work. However, this also threatens to destroy still untouched natural resources through urban sprawl.

The views of the experts

Are these three scenarios the options for Europe in 2020? There was a general agreement among the respondents that the scenarios covered the main options, although some respondents saw relatively few conflicts between these options. For example, 'the three scenarios do not really contrast. In my opinion, they illustrate rather different aspects of one scenario'. This comment highlights the extent to which the growth scenario has a business focus, the equity scenario a household focus, while the environment scenario represents an alternative life-style. Despite the high degree of consensus among respondents regarding these scenarios, a number of interesting modifications were suggested, particularly with respect to the substitution of travel by telecommunications and the spatial consequences of telecommunications.

The substitution of travel by telecommunications

A continued growth in communication services and quality is anticipated by most respondents. To some, this is a natural evolution and the growth scenario will come automatically. Others take the optimistic stance that their preferred scenario combining growth, equity and environment elements will come naturally. In this case, however, the critical question concerns the extent to which telecommunications will provide a substitute for travel.

The optimistic view is summarized by a respondent from France, 'Stimulate substitution for travel by modern "media". Improve education levels and the quality of TV will follow fast. Introduce a day without TV to avoid overfeed reactions (a day a month may be okay because nobody will accept a day a week). TV is already closer to our hearts than the motor car'. However, many other respondents expressed their concern about the tendency of telecommunications to generate travel rather than replace it.

In some experts' opinions, there is a considerable risk that both transport and communications will continue to grow to form a 'high-info-mobility society'. To avoid this outcome telecommunications costs must come down drastically to limit mobility. 'Increase prices for travel and give greater advantage to telecom opportunities' is one recommendation.

An important question which will have a profound influence on the degree to which telecommunications substitute for travel is the extent of government control of telecommunications. Although some respondents felt that there should be a free market in telecommunications, it was also argued that the institutional rules governing the telecommunications sector must be changed. Government investment levels should be kept high and subsidies provided for peripheral regions.

The spatial consequences of telecommunications

It was generally agreed that, in principle, there seems to be no major obstacle to the continued dispersion of high quality telecommunications. Consequently, telecommunication services can be used as a means of decentralization. As a British respondent pointed out, 'better communications promote dispersed work, with Ireland and Scotland as good examples of regions that have attracted service work from the English cities'. Alternatively, in the words of a Danish commentator, 'if it is the needs of business which decide, I don't think they will favour central regions'.

While telecommunications provide a means for spatial decentralization in theory, there was no general agreement as to the extent to which they will be utilized for this purpose. It can be argued that 'the decentralization scenarios presented are predicated on the concentration effect being relatively short-lived, before advanced infrastructures and services diffuse universally'. Nevertheless, as a French respondent pointed out, 'what has to be looked for is transparency of data transmission through the national borders. This is a prerequisite of EDI (Electronic Data Interchange) but is not good enough in Europe because of the national monopolies of the PTTs'. Furthermore, as a German respondent notes, 'governments should not leave the telecommunications field to the market but take an active role in guaranteeing equitable access to the media. Semi-public forms of media such as the radio and TV systems in Britain and Germany should be promoted. However, the existence of democratic control is paramount'.

The policy implications of these viewpoints are considerable. In the words of one expert:

> The equity and environment scenarios are variants of a single scenario, one which stresses the decentralizing potential of advances in telecommunications. This potential clearly exists, and undoubtedly constitutes a plausible, and highly necessary, objective for policy development. One of the big question marks over the telecommunications field, however, concerns the regulatory structure within which advanced services are developed and delivered. Although the background discussion and the growth scenario description both note the way in which industrial and commercial demands are leading

the diffusion process, more explicit mention needs to be made of the liberalization of telecommunications markets.

The main issues involved in the debate regarding the spatial consequences of current developments in telecommunications in Europe were summarized by a British respondent:

> The essential characteristic of telecommunications in a liberalized market-driven environment is that it poses a fundamental challenge to the whole notion of universal service provision. For advanced infrastructures, such as those necessary to support broadband ISDN, the market threshold may simply not be reached to justify the investments (which, in spite of falling prices are likely to remain substantial). The risk here for certain locations/types of area is not therefore only about *when* they will be wired up, but *whether* they will be wire up at all.

Which scenario?

The respondents gave the growth scenario very strong support. About 60 per cent (35) saw it as the most likely, while the equity scenario ranked second (17), and the environment scenario last (2). As to the most preferred scenario, however, almost half of the respondents voted for the equity scenario, with the growth scenario second (11) and the environment scenario third (9).

Speculation

In the future we can see more effective logistics, smarter cars on 'intelligent' roads as well as 'intelligent' houses that are responsive to the demands of their users (see Text Box 11.1). Microelectronics technology comes in gradually in most products of our everyday life. Office computers communicate with each other and with home-based terminals. Satellite mounted sensors monitor the conditions of the earth. Other communication satellites relay messages from trucks in remote areas to their home base. Radio technology invades local telephone networks. Mobile telephones are connected by cellular radio to high capacity ISDN networks. Communication becomes mobile. Personal communication is possible

A MORNING AT XANADU

6.30am As the sun begins to rise, the analytical left side of Xanadu's House Brain computer system comes to life. A quick scan through its memory banks shows that the automatic systems operated during the night by its more reflexive right side (without human command or reprogramming) are all functioning properly. Every six minutes around the clock, the computer monitors its sensors to confirm that water pressure is at correct levels in the plumbing system, the solar sinks hold adequate heat reserves, the hydroponic garden greenhouse has the correct levels of moisture and nutrients, all electronic circuits are functioning, all windows and doors are secure, no structural faults or leaks have developed and no high-priority messages have been received since the previous status check. If anything had gone wrong, the appropriate sensors would have alerted the House Brain's analytical circuits and fed them the information available to decide whether the problem could be handled by computer or if it warranted waking someone for instructions.

Now comes the most important task of all—checking each of the house's still-sleeping human residents to make sure that they too are all 'functioning correctly'. Using sensors built into each bed, the House Brain can measure body temperature, brain activity and pulse rate.

Next, the House Brain tunes in to the outside world. Using the outside sensors, the temperature, humidity and barometric pressure are checked and compared with its memory banks to detect the possibility of upcoming storms or other radical weather changes.

The House Brain now searches the videotex system for items that the master and mistress of the house have preprogrammed under key words such as 'real estate', 'stock market' and specific company names. Once found, these items are separately stored for retrieval on request.

7.15am Slowly, the temperature on the mistress's side of the bed begins to rise and a gentle massage, like thousands of tiny fingers, becomes more obvious. Since the master does not have to wake up for another 45 minutes, the Brain has decided it should communicate today's messages for the mistress using only text instead of voice synthesis. 'Good morning. It is now 7.16. Your appointment in Tampa is at 9.30 but since traffic is light today you will not have to leave until 8.15 if you take Route 1-4.'

8.00am As the bed warms up and begins to vibrate vigorously, the House Brain's voice wakes the master of the house. 'Time to wake up, John. It's 8.01. You still have to do your morning exercises and make sure Johnny has his breakfast before he starts his school work. It's a beautiful day and you must be at your best for your teleconference. You have two messages, but none with a high priority. I have stored three articles for you to read; one of them is about your company. As soon as you are ready, I will retrieve them for you. Now you really must get up!'

Text Box 11.1 A morning at Xanadu (Mason et al., 1984). The Xanadu house is one of several prototype houses that have been built to demonstrate the potential of architronics for family life in the late 1990s. This imaginary record of a typical morning in such a house in the late 1990s graphically illustrates the options that are already available as a result of recent developments in telecommunications (see also Mason et al.,1983).

anywhere and any time. The limiting factor is imagination. It will take some time before we can see these services as part of our own lives.

As there seem to be no specific negative effects of telecommunication services, environmentally or otherwise, growth in telecommunications seems to be acceptable. However, for equity reasons care has to be taken that telecommunication services are made available equally to all groups of society in cities and rural regions alike. Unfortunately, nothing guarantees that this will be the case in most countries. Rather, in a deregulated telecommunications market, the chances are that the more advanced services will only be affordable for businesses and affluent individuals, with the effect that the 'information society' will be divided into information-rich and information-poor, a new sort of social cleavage. In addition, it is likely that without strong government intervention the most advanced telecommunications services will become available first, or only, in the largest agglomerations and so further reinforce their dominance over the peripheral regions.

Between these technological dreams and fears lies the reality of policy-making. Telecommunications policy is, like other policy fields, caught between global competition, the desire for European integration and coordination, powerful industry interests and the concerns of domestic labour unions, regional representatives and social critics. In this situation, policy-makers have in principle two options:

either   to continue to promote the demand-driven concentration of high level telecommunication infrastructure in core regions—the *laissez-faire* option with industry and business as leaders of the telecommunications development—and to proceed with the deregulation of telecommunication services resulting in insufficient services for businesses in peripheral low-demand regions;

or   to promote telecommunications in peripheral regions by subsidizing investments for telecommunications infrastructure and introducing flat rate (not distance-dependent) telecommunication charges—this is the decentralization option available to dampen the growth of the largest agglomerations by establishing equivalent service, culture and information work opportunities in small cities and peripheral regions.

The discussion about the direction of a long-range telecommunications policy for Europe has not even started. The European Commission, in its traditional perspective fixed to economic growth, sees telecommunications predominantly as a prerequisite for successful competition on a global scale with the USA and Japan. The national governments follow their own, uncoordinated telecommunications policies largely dictated by the interests of their national telecommunication industries or pursue a *laissez-faire* policy of deregulation. The failure to develop a clear European vision of which telecommunication policy is best for Europe and hence which infrastructures and services need to be implemented, may in the long run turn out to be a costly mistake.

Further reading

Telecommunications are a very recent but rapidly expanding field of scientific inquiry. A classic in the field of communications is the book by Masuda (1981) on the Information Society. Remarkably, but quite appropriately, the distinction between telecommunications technology and information technology in general tends to become blurred. Research on telecommunications and/or information technology in general focuses on three aspects.

The first group of studies looks into the diffusion of telecommunication and information technologies in different sectors or countries. Two important documents discuss issues of trans-border telecommunication networks: Ungerer and Costello (1988) in a report for the European Commission describe the prospects for the telecommunication market in Europe after the completion of the Single European

Market in 1993, while the report of the OECD (1989) discusses policy implications of network-based telecommunication services on a global basis.

A second group of studies is interested in the impacts of telecommunication and information technology on life-styles and production patterns in specific industries or services. For instance, Miles (1988) reports on the diffusion of information technology in households and the resulting changes in communication behaviour, Blackburn et al. (1985) the impacts of technological change on manufacturing and Moss (1987) the impacts of telecommunication on international financial centres. As already mentioned in the previous chapter, Snickars (1987) and Salomon (1988) discuss the relationship between transport and telecommunications.

The third and largest group investigates the spatial impacts of telecommunications and information technology. In the majority of cases telecommunications and information technology is seen as an, albeit essential, part of a larger package of technologies including, among others, robotization, logistics, but also special product technologies such as semi-conductor technology or biotechnology. This general concept of high-tech is used, for instance, by Thwaites and Oakey (1985) or Aydalot and Keeble (1988) or in some of the references given in the chapter on regional development (Chapter 7). Other authors look at the regional impacts of telecommunication and information technology together, for instance Mackintosh (1986), Orishimo et al. (1988), Giaoutzi and Nijkamp (1988), Robins and Hepworth (1988), Hepworth (1989) and some contributions in Brotchie et al. (1990). There is only little work as yet that explicitly addresses the spatial impacts of telecommunications. Examples are the chapters by Steinle and by Goddard and Gillespie in Giaoutzi and Nijkamp (1988).

Part III
Choices for Europe

Chapters 3 to 11 have revealed a great variety of trends, opinions and policy options. It seems that for each of the nine fields, the perceptions, expectations and preferences extend over such a wide range of perspectives and policy orientations that a *consensus* about likely—let alone desirable—future developments seems almost impossible. If this is true for a relatively homogeneous group of experts, how can a consensus about future policy options be achieved in a Europe of different nations, levels of affluence and cultural attitudes or even within individual countries with different regions, interest groups or political fractions and their unmeasurable diversity of attitudes, beliefs, hopes and aspirations.

This book does not claim to show a straight avenue towards consensus. All it can attempt is to provide an as systematic as possible overview on the basic options available for transport and communications policy. None of these options will be optimal in all fields concerned; the achievement of one goal tends to be at the expense of others. So to present options for choice also means naming the price that has to be paid if they are chosen, and that means controversy.

As it will be seen, the controversy can almost universally be represented in the context of the growth-equity-environment paradigm. In which direction will, or should, a future Europe as a world region move? Rivalry with economic competitors in East and West suggests a strategy for more growth in order to persist in the contest for new or established markets. So do the still grave disparities in affluence between the regions in Europe. However, every growth policy unwittingly carries the risk of aggravating the differences in well-being it is meant to reduce, if on a higher level. A strict egalitarian regional policy, however, may imply a loss of growth, or at least that is claimed by the proponents of growth strategies. Irrespective of this controversy, can further growth be defended for a region that already today, together with North America, consumes an outrageously high percentage of the world's energy resources? The advocates of the environment say 'no', and in the context of transport and communications this means also a strict 'no' to the seemingly relentless trend to ever more mobility. The implications are profound because mobility has almost become a synonym for modern life. To halt the further growth of mobility, or even to put the clock back, would require fundamental changes in our way of life, our cities and the pattern of production and distribution. In most cases it implies a loss of potential interaction, pleasure and above all convenience. Moreover, without accompanying redistributional safeguards, constraints on mobility almost inevitably contain an element of inequity as the well-to-do will be able to buy themselves a disproportionate share, and this is but one example in which environmentalist strategies conflict not only with growth but also with equity concerns.

The main objective of Part III of the book is to clarify some of the interdependencies between the options for choice in transport and communications for decision-makers in Europe. It is a synthesis of the previous chapters, which draws upon the results of the nine component scenarios but also seeks to provide a *cross-cutting*, holistic perspective which pulls the arguments put forth in them together and points to the cross-impacts, synergies and conflicts between them.

Part III consists of two chapters. Chapter 12 examines what the experts in the panel group involved in the study suggest as guidelines for future policy-making in Europe and in their respective countries. Chapter 13, finally, is left for the authors' summing up and synthesis, together with some speculation and, perhaps, a grain of optimistic imagination.

CHAPTER 12

TOWARDS A NEW PARADIGM

Before a synthesis of the findings of the analysis is attempted in the final chapter, the experts participating in the exercise will again be consulted. However, this time their specific responses to particular issues and problems are not sought. Instead how they respond as a *body* of experts, representing a cross-section of academics and professionals in no less than eighteen countries, is examined.

It is important to stress that the selection of participants did not conform to strict standards of statistical representativeness but was basically determined by membership in NECTAR, a European research network in the field of transport and communications initiated by the European Science Foundation (see Chapter 1). Nevertheless the respondents are a fair cross-section, if not of the *population* of the participating countries, then of their community of researchers in the fields of transport and communications and, as this is a field not dominated by one particular discipline, through a wide range of academic backgrounds and professional affiliations. At one point during the study the panel was enlarged by the inclusion of a number of younger scientists because it was felt that people under thirty were underrepresented, and in this case NECTAR membership was overridden as a selection criterion.

The image of transport planners

Taken as a group, the transport planning community today has a doubtful reputation. The times have long passed when railway engineers and tunnellers were the heroes of industrial progress and technological efficiency. Today the profession is much more associated—rightly or wrongly—with the planning disasters of the recent past. After all, transport engineers were responsible for the atrocious elevated motorway superstructures encroaching on many cities, the *tabula rasa* planning for the sake of undisturbed traffic flows, the enormous waste of urban and rural land for highways and airports, traffic congestion, noise and pollution and all the rest of the evils connected with modern transport. So the image of the profession is one of stubborn technocrats who in their narrow-minded rationality always place technical efficiency over less tangible concerns such as environmental or social considerations.

As it turned out, the survey provided an interesting opportunity to examine whether this familiar prejudice about the transport planning community is justified. This was made possible by two standardized questions in an otherwise unstandardized questionnaire. While the non-standardized responses produced most of the points made in the preceding chapters, these two questions allowed the aggregation and grouping of the respondents in quantitative terms.

The two standardized questions asked which of the seed scenarios were considered to be (a) most likely and (b) most desirable from the perspective of the country of the respondent. In other words, the first question revealed how the respondents *viewed* reality and the second how they *wished* reality should be. The difference between the two can be used as a *measure of satisfaction* with the way things are: if the scenario seen as most likely is also the one seen as most desirable, satisfaction is perfect, conversely if the most likely scenario is also the least desirable, satisfaction is nil. To facilitate comparisons, perfect satisfaction is given the value of one hundred and nil satisfaction is given the value of zero, and the resulting index is called the *index of satisfaction*.[1]

If indeed transport planners are narrow-minded technocrats as current prejudices claim, they should be quite happy with the world they have helped to create, i.e. they should show a high index of satisfaction. If anything they should regret that transport improvement is too slow and wish that transport policy be more growth-oriented. Is this hypothesis borne out by the results?

Perceptions and attitudes

Figure 12.1 gives the answer. Here the responses are aggregated by discipline in a triangular coordinate space the corners of which are associated with the three overall goals, growth (A), equity (B) and environment (C). Each response, or group of responses, can be located in this coordinate space as a pair of points indicating the 'most likely' and 'most preferred' scenarios, respectively.[2] In this case the 'most likely' scenario is indicated by a hollow circle and the 'most preferred' scenario by a solid circle. The figures show that the average satisfaction index of transport planners with an engineering background is higher than those of the other disciplines. In

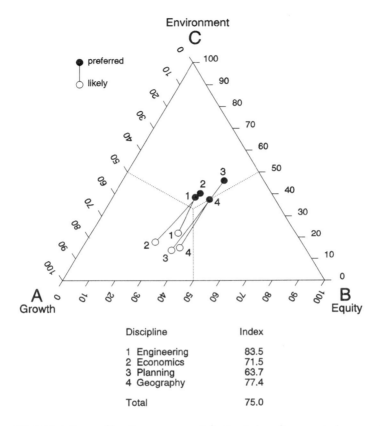

Discipline	Index
1 Engineering	83.5
2 Economics	71.5
3 Planning	63.7
4 Geography	77.4
Total	75.0

Figure 12.1 Discipline profiles. The average satisfaction index of transport planners with an engineering background is higher than those of the other disciplines. In particular urban and regional planners are much less in agreement with how things develop in their field. All disciplines agree that the trend in transport and communications points in the direction of growth rather than equity or environment, but feel that a shift away from a growth-oriented transport policy would be desirable. The desired shift clearly points towards the environment, not towards equity.

contrast urban and regional planners are much less in agreement with how things develop in their field. However the figure contains a surprise: all disciplines agree that the trend in transport and communications points in the direction of growth rather than towards equity or environment, but in contrast to the prejudice quoted above they feel that a shift away from a growth-oriented transport policy would be desirable. The desired shift clearly points towards the environment, not towards equity. Even the engineers among the respondents would prefer a more environment-conscious transport and communications policy!

Is this an artefact of the composition of the sample group? Figure 12.2

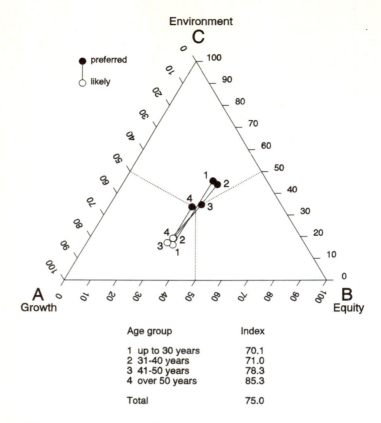

Figure 12.2 Age-group profiles. All age groups think that transport planning should be less growth-oriented. The differences between the age groups are in line with common wisdom about age-group behaviour: the younger experts are more radical than their elder colleagues; they are less satisfied with current trends and generally are more concerned about environmental questions. The middle-aged and senior experts are less critical about the trends as they are, and this is no surprise as this generation was actively involved in establishing them. However, even they think that things have to change towards a better environment.

groups the respondents not by discipline but by age. A different picture emerges but the basic pattern is the same: all age groups think that transport planning should be less growth-oriented. The differences between the age groups are in line with conventional wisdom about age-group behaviour: the younger experts are more radical than their elder colleagues; they are less satisfied with current trends and generally more concerned about environmental questions. The middle-aged and senior experts are less critical about the trends, and this is no surprise as this generation was actively involved in establishing them. However, even they think that things have to change towards a better environment.

There are also significant differences between how men and women see current trends in transport and communications (see Figure 12.3). Women are much more critical about current transport planning than men, in fact their satisfaction index is the lowest of all groups analysed. They feel even more strongly than men that environmental concerns tend to be neglected in current planning processes and that they should be given much more weight. If only equity considerations are examined, women are less concerned than their male colleagues. However, these results must be interpreted with caution as women were greatly underrepresented in the sample (see Figure 1.2) and without exception belonged to the youngest age group.

If in all the above cases the respondents voted for less growth and more

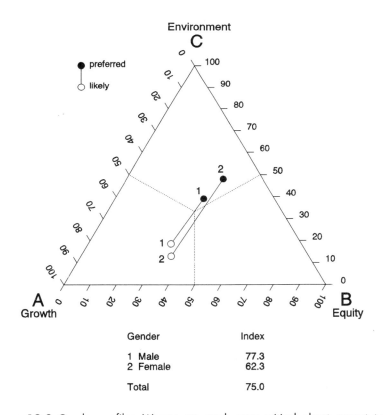

Gender	Index
1 Male	77.3
2 Female	62.3
Total	75.0

Figure 12.3 Gender profiles. Women are much more critical about current transport planning than men, in fact their satisfaction index is the lowest of all groups analysed. They feel even more strongly than men that environmental concerns tend to be neglected in current planning processes and they should be given much more weight. If only equity considerations are examined, women are less concerned than their male colleagues. However, these results must be interpreted with caution as women were greatly underrepresented in the sample and without exception belonged to the youngest age group.

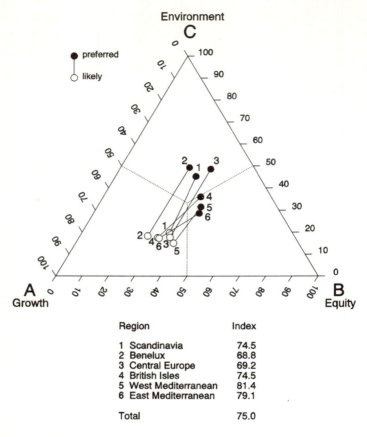

Figure 12.4 Region profiles.

Region	Index
1 Scandinavia	74.5
2 Benelux	68.8
3 Central Europe	69.2
4 British Isles	74.5
5 West Mediterranean	81.4
6 East Mediterranean	79.1
Total	75.0

Figure 12.4 Region profiles. Respondents from northern countries were the least satisfied and the most environmentally concerned. Their colleagues from Mediterranean countries, though some of them are likely to favour the expansion of the transport and communication infrastructure in their countries, unanimously think that protecting the environment is even more important. Interestingly, their preferred scenario is closer to equity than to environment because they know about the disadvantage of being poorly connected.

environmental policies in transport planning, the situation should become more diversified if individual countries are examined. After all, the countries in Europe are very different with respect to their transport and communications infrastructure. Some countries, in particular the industrialized countries of central, north-western and northern Europe, have highly developed transport and communications networks, while there are still serious deficiencies in infrastructure in countries like Portugal, Spain, southern Italy, Greece and Turkey. Accordingly, though there may exist a certain saturation effect in the northern countries, at least the Mediterranean countries should be expected to push for *more* rather than less growth.

TOWARDS A NEW PARADIGM

However, as Figure 12.4 shows, this is not the case. This figure shows likely and preferred scenarios for six regions or groups of countries.[3] It is true that respondents from northern countries were the least satisfied and the most environmentally concerned, with Benelux and central Europe having the lowest satisfaction index. However, their colleagues from Mediterranean countries, though some of them are likely to favour the expansion of the transport and communication infrastructure in their countries, think that protecting the environment is even more important. Interestingly, their preferred scenario is closer to equity than to environment because they know about the disadvantage of being poorly connected. The respondents from the British Isles occupy a middle position. It may seem surprising that the respondents from Mediterranean countries have the highest satisfaction index though clearly their infrastructure is less developed. The high index in this case merely indicates that they are less in opposition to the prevailing growth-oriented policy in their countries and wish only to make minor adjustments to it.

So whatever grouping of the respondents one chooses, the result is unequivocal. There is an overwhelming consensus that the current growth-orientation in transport planning is harmful and should be replaced by a more environment-conscious or more equity-oriented kind of policy. The kind of policy shift advocated can be made more explicit by dividing the triangle used in Figures 12.1 to 12.4 into three areas. In each of these areas one of the three goals is dominant. Now the respondents can be classified by type of policy shift advocated. Figure 12.5 visualizes the result. It can be seen that for the majority current transport planning is growth-oriented but should be more oriented towards environmental issues. A much smaller proportion would like to see a policy shift from growth to equity, and only relatively few advocate a shift from equity to environment. Only a minority of the respondents demand no policy shift.

Shift of Paradigm?

This result is encouraging. It says no less than that the familiar prejudice that transport planners are narrow-minded advocates of growth does not hold true—if the sample of experts chosen for this exercise is at all representative of transport planners in the countries included in the study. If this is the case—and there is no serious reason to doubt it—this result indicates a *major shift of paradigm* in the transport planning community. The importance of this shift cannot be overestimated. If indeed the experts concerned with the design and implementation of transport infrastructure develop a more comprehensive set of values with respect to the purpose and objectives of transport planning, this cannot fail to have its

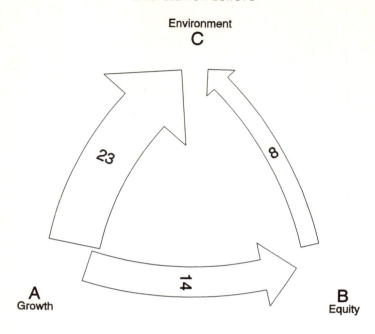

Figure 12.5 Advocated policy shifts. The respondents can be classified by type of policy shift advocated. It can be seen that for the majority current transport planning is growth-oriented but should be more oriented towards environmental issues. A much smaller proportion would like to see a policy shift from growth to equity, and only relatively few advocate a shift from equity to environment. Only a minority of the respondents demands no policy shift.

impacts on the public discussion on transport matters even in countries where the public opinion is still less advanced.

One note of caution seems appropriate. It may well be that the shift in paradigm is not so fundamental as it appears and should be more seen as an *adjustment* to a changing professional environment. Several comments of the respondents reveal that there is a widely held view, especially among respondents with engineering or economics backgrounds, that without growth neither equity nor a clean environment can be achieved, but that growth, if appropriately managed, can reduce disparities and also support environmental improvement. The belief that the conflicts between efficiency and equity and between economy and environment can be resolved should certainly not be lightly dismissed, but it may prove to be harmful if it is misused merely to defend the continuation of business-as-usual growth-oriented policy-making.

Problem areas and policy fields

One disadvantage of the present aggregate analysis is that the concepts of 'environment' or 'equity' remain rather loosely defined. What is meant by an 'environment-conscious' or an 'equity-oriented' policy, and do the numerical responses say anything about possible policies? Figure 12.6, in which the responses are grouped by component scenario, may give some hints. Component fields in which the discrepancy between the likely and preferred trend is large are obviously priority candidates for policy action; conversely where this discrepancy is small, the need (or possibility) for policy action may be less. Five component fields stand out by having a very low satisfaction index:

2 *Life-styles*
5 *Regional development*
6 *Urban and rural form*
7 *Goods transport*
8 *Passenger transport*

These are the fields in which the respondents see the greatest need and potential for change. In all five fields environmental considerations are felt to deserve more attention. This is particularly true for *Goods transport* and *Passenger transport* where the largest improvements for the environment can be achieved. The other three fields have also a strong social dimension which means that a compromise solution between equity and environment needs to be found.

It is interesting to note that not only *Goods transport* and *Passenger transport* are suggested as policy areas but that these are recognized as *derived* activities resulting from the temporal and spatial organization of society. *Economy* should have been listed here too, because goods transport largely depends on the spatial organization of production and distribution. It seems that the respondents found it easier to criticize the consequences of growth than growth itself which is, after all, the origin of material wealth. Somewhat surprising, too, that *Environment* itself did not feature as a priority policy area. However this can be explained by the fact that the questionnaire only addressed transport-related environmental impacts, and these can indeed be reduced only by transport-related policies. *Population* received a relatively high satisfaction index indicating that population development was not seen as critical nor as a prime policy area.

This is about as far as a *numerical* analysis of the responses can be carried. It does not say anything about the specific policies suggested by the experts for each field. For this the written comments of the respondents need to be used. This will be done in the final chapter.

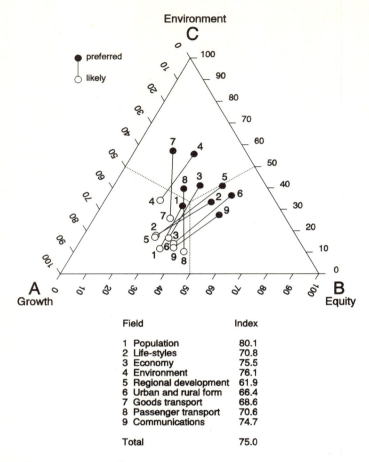

Field	Index
1 Population	80.1
2 Life-styles	70.8
3 Economy	75.5
4 Environment	76.1
5 Regional development	61.9
6 Urban and rural form	66.4
7 Goods transport	68.6
8 Passenger transport	70.6
9 Communications	74.7
Total	75.0

Figure 12.6 Component profiles. Five component fields stand out as having a very low satisfaction index: 2 (*Life-styles*), 5 (*Regional development*), 6 (*Urban and rural form*), 7 (*Goods transport*) and 8 (*Passenger transport*).These are the fields in which the respondents see the greatest need and potential for change. In all five fields environmental considerations are felt to deserve more attention. This is particularly true for *Goods transport* and *Passenger transport* where the largest improvements for the environment can be achieved. The other three fields have also a strong social dimension which means that a compromise solution between equity and environment needs to be found.

Notes

1. Altogether sixty respondents participated in the exercise and rank-ordered the three seed scenarios for each of the nine fields. The resulting 540 votes established the data base of the following analysis. The satisfaction index S is calculated as follows:

$$S = 100 - \{abs[w(r_a)-w(q_a)]+abs[w(r_b)-w(q_b)]+abs[w(r_c)-w(q_c)]\}$$

where r and q are the ranks of the scenarios with respect to likelihood and desirability, respectively, and subscripts $_a$, $_b$, $_c$ refer to component scenarios A (growth), B (equity) and C (environment), respectively. The weights w indicate the importance of the rank order in terms of satisfaction. In the following analysis the first rank was given a weight of 67 per cent, the second rank 33 per cent and the third rank was disregarded.

2. Using the terminology of Note 1, the coordinates of the most likely scenario are $w(r_a)$, $w(r_b)$ and $w(r_c)$, while the coordinates of the most prefered scenario are $w(q_a)$, $w(q_b)$ and $w(q_c)$.

3. The six regions are: Scandanavia (Denmark, Norway, Sweden), Benelux (Belgium, Netherlands), central Europe (Germany, Switzerland, Austria), British Isles (United Kingdom, Ireland), west Mediterranean (France, Portugal, Spain) and east Mediterranean (Italy, Greece, Yugoslavia, Turkey).

CHAPTER 13

ISSUES FOR DECISION-MAKERS

Chapters 3 to 11 of this book have presented trends and scenarios for the nine fields of population, life-styles, economy, environment, regional development, urban and rural form, goods transport, passenger transport and telecommunications. Yet the real world cannot be compartmentalized into categories. The nine fields interact in a complex, dynamic way, and it is necessary to consider them together. This final chapter attempts a synthesis and presents a number of choices for policy-makers.

The chapter consists of four sections. In the first section the trends and tendencies observed in the nine fields will be related to each other as the joint outcome of a few fundamental 'megatrends' influencing our society. The second section summarizes the most likely future development of these trends in a holistic scenario as seen by the experts participating in the survey. Section three confronts this most likely holistic scenario with the scenarios preferred by the participants. This confrontation leads to the fourth and final section, in which choices for decision-makers at the regional, national and European levels will be discussed.

Cross-cutting trends

The trends and tendencies observed in the nine fields discussed in Chapters 3 to 11 display an overwhelming panorama of simultaneous, partly connected and partly unrelated, partly mutually reinforcing and partly contradictory, developments. The question immediately coming to mind is whether there is a common theme, a joint origin that binds them together.

The authors' conviction is that it is indeed possible to explain the seemingly disparate trends shaping modern societies in the context of broad movements related to the three paradigms of *growth*, *equity* and *environment*.

Growth

The growth paradigm, the oldest of the three, incessantly drives countries and regions, groups and individuals in a never-ending competition for wealth, power and happiness. The growth paradigm is also deeply related to the idea of progress. The major part of the changes in industrial societies can be related to technological progress. The incessant flow of technological innovation brought about by ubiquitous competition in market economies is at the heart of the growth of production and income in western Europe since the Second World War. Innovation has been the prime force behind the restructuring of economies characterized by declining employment in the primary and secondary sectors and a concurrent growth in the service economy and the shift in labour demand from low-skilled to highly skilled occupations. The overall effect is the gradual disappearance of heavy physical labour and a general reduction of work hours plus an unprecedented diversity of products and services offered.

Innovation in medicine has helped to reduce mortality, rising incomes and improved social security have reduced the need for high fertility, so populations tend to stagnate or decline, households tend to become smaller, and the proportion of old people tends to grow. Declining work hours mean more leisure time and this gives rise to both mass culture and an increasing variety of life-styles served by a growing diversity of opportunities not only for recreation and entertainment but also for education and culture. The growing range of tastes and preferences creates the demand for the diversity of products and services stimulating the transformation of the economy.

Technological progress is also responsible for the vast increase in mobility of persons and goods. The aircraft and the car have revolutionized mass passenger travel and opened up a new world of spatial independence and freedom of choice, of human exchange and cultural encounters. Goods transport by lorry has connected even the remote parts of the European continent to the urban centres and has greatly contributed to the internationalization of production and markets which can be seen in the great variety of goods from far-away countries in our stores.

Last, but not least, technological innovation has led to the explosive growth of telecommunications and information technology that has revolutionized every facet of our daily lives, by fundamentally transforming production and distribution through the logistics revolution and globalizing national economies as well as the worlds of finance, science, politics, science and the arts.

In summary, the pursuit of growth has led to an unprecedented expansion, diversification and intensification of production and consumption and exchange and communication processes coupled with an ever growing degree of complexity, interconnectedness and interdependence of sub-systems. The growth paradigm has also fundamentally transformed the relationship between time and space. Increasing density and globalization of interaction has accelerated adjustment processes and the diffusion of knowledge and cultural values. New high-speed transport modes and telecommunication have reduced the barriers of distance and stimulated the dispersal of human activities, the spatial division of labour in the economy and interregional and international cooperation and exchange. Growth is the essence of the well-being of industrialized societies. When it is absent during an economic recession, a feeling of crisis and insecurity prevails.

Equity

Both the equity and environment paradigms emerged when fundamental limits and shortcomings of the growth paradigm became visible. The equity paradigm is the older of the two. It dates back to whenever people revolted against a suppressive ruling class but it has become a major societal force since the French Revolution. Egalitarian movements were not opposed originally to technological progress but turned into critics when it became apparent that growth-oriented policies tended to increase social inequality. On a global scale, the notion that the growth enjoyed by the industrialized countries was only possible at the expense of the poor countries of the Third World was recognized during the post-war years when the colonial powers retreated from their possessions. On a more domestic scale it became more and more apparent that the unleashed forces of market competition did indeed lead to greater overall affluence but that this affluence tended to be less and less evenly distributed both across the population *and* across the countries of Europe, the regions of a country or the parts of a region.

Today it is generally recognized that economic growth and growth-oriented strategies tend to increase social and spatial polarization unless strong counteracting policies are introduced. Migration, such as continuing rural-to-urban migration in peripheral regions or interregional labour migration from peripheral to central regions or from Third World countries into the European Community, reduces regional disparities but not enough to equalize living conditions between their origins and destinations.

There is also agreement that transport and communications infrastructure, in particular high-speed transport and advanced telecommunications, tends to increase growth opportunities disproportionately at the central locations already served best by the existing infrastructure and so contributes to spatial polarization rather than equalizing accessibility. It is also argued that high-speed transport, despite mass air tourism, basically caters to the excessive mobility demands of a privileged minority, whereas the majority of daily commuters suffer from degraded local public transport. By the same token the private car is increasingly criticized because, given the deplorable state of public transport, it divides society into the mobility-rich and the mobility-poor that are too young, too old or too poor to drive a car.

Environment

The environment paradigm is the youngest of the three. It originated from the growing awareness that the natural resources of the globe are limited and that economic growth is approaching these limits or has in many respects already exceeded them. From the beginning, the environment paradigm was opposed to the growth paradigm. Its relationship to the equity paradigm is less clear, particularly as its supporters tend to be the same people as the ones fighting for social justice. It is only relatively recently that conflicts between environmental and equity goals have appeared.

The main thrust of the environment paradigm is against continued growth. Especially in the countries with the highest levels of production there is increasing resistance against the further depletion of natural resources for economic objectives, even though in these countries the efforts to restore the environment after decades of neglect are most advanced. Conditions tend to deteriorate in the countries of Europe with declining industries, where industrialization is still in its early phase or in the industrial areas of eastern Europe. Here opposition against industrial waste and pollution is still weak because jobs are considered more important than the environment. In these regions environmental concerns are not only in conflict with growth but also with equity goals.

The conflict between the growth and environment paradigms is nowhere so pronounced as in transport. New airports, motorways and high-speed rail lines are the symbols of growth which consume large amounts of virgin land in environmentally sensitive areas and agglomerations and are the cause of a great deal of air pollution and noise intrusion. Consequently opposition against such projects is commonplace and increasingly presents a serious obstacle for large infrastructure improvements. In many countries the time needed between the first conceptualization and the opening of simple road projects can be as long as ten to fifteen years.

However, the opposition to large transport projects more fundamentally challenges the general trend to more mobility and a greater spatial division of labour. The large investments in high-speed transport must be compared with the neglect and degradation of local public transport and the concomitant growth in private car ownership which leads to ever more urban sprawl, pollution and noise and road congestion and a substantial toll of deaths and injuries in road accidents. This contrasts with the ideal of sustainable urban development with constrained energy use, short distances and a reduced need for mobility using more environment-friendly transport modes. The growing movement against the destruction of the European environment under the dominance of the car is one of the strongest trends in urban and transport planning in Europe today.

The most likely scenario

Growth, equity and environment are three partially conflicting paradigms that will influence the future geography of Europe. Which of them will be the most powerful? In this section a holistic scenario of the spatial development of Europe will be presented that, in the opinion of the experts participating in the scenario experiment, has the highest degree of probability of becoming reality. This 'most likely' scenario will be assembled from the component scenarios presented in the previous chapters.

The overwhelming evidence of the experiment is that the growth scenario is by far the most likely. The majority of the respondents believed that if present trends continue the market economies of western Europe will continue on their growth path. The essential characteristics of the growth scenario, as assembled from the component scenarios, are set out on the left-hand side of Table 13.1.

— *Population*. In the year 2020 one quarter of the population in central Europe is over sixty. Growing life expectancy and improved health have enabled elderly people to participate in sports, leisure and travel longer. Only continued growth in productivity makes it possible for the shrinking number of economically active to support the growing number of pensioners. The government has relaxed immigration restrictions for workers from developing countries without granting them permanent European citizenship.

— *Life-styles*. Singles and 'dinks' are the engines of the growth economy of the early twenty-first century. The economically independent young white-collar worker is the model for the life-style of the decade, characterized by efficiency, high mobility, intensive use of telecommunications and vast consumption of energy and resources. There is a massive growth of commercial leisure activities. However, the success of the young and active rests on the declassification of the less able and efficient.

Table 13.1 Summary of likely and preferred scenarios

Field	Most likely scenario	Most preferred scenario
Population	Low birth rates, ageing society; growth-financed social security; non-EC foreign labour without citizenship.	Crisis of social security system overcome by immigration from developing countries; government support of young families.
Life-styles	Singles and 'dinks' model for life-style: efficiency, mobility, telecommunication, consumption; declassification of less able.	Change of values: solidarity instead of competition; renaissance of the family; participation in community affairs emphasized.
Economy	'Fortress Europe' economic empire; income disparities between European core and periphery and within European countries.	European government promotes sustainable development; taxes on luxury goods, rigorous emission standards; alternative technologies.
Environment	Serious congestion and transport-generated pollution; massive land consumption for new motorways, high-speed rail lines and airports.	Europe leader in environment-conscious policy-making; use of fossil fuels constant; heavy taxes on car ownership and petrol; public transport growing.
Regional development	Further concentration of economic activities in the European core; severe agglomeration diseconomies; economic decline in peripheral regions.	Decentralization programmes and strict land-use control in urban areas; incentives for location in peripheral areas; decentralization of transport infrastructure.
Urban and rural form	Growing spatial segregation of social groups in cities; suburbanization of manufacturing; disappearing countryside.	Disincentives for location in large cities; financial aid for small cities; land speculation curbed; car restraint policies.
Goods transport	Dramatic increase in road freight transport, toll motorways and bridges; rail freight service disappeared.	Restriction and taxation of road freight transport; air freight regulated; promotion of ecological vehicles for local distribution.
Passenger transport	Highly mobile society; dominance of individual automobility; local public transport declining; competition between car, high-speed rail and air.	Car use constrained, renaissance of public transport; clean cars provide harmless mobility for dispersed society.

195

Communications	Massive use of fibre optics and satellite communications; 'information city' changes life-styles; dominance of large cities reinforced.	Use of telecommunications for equalizing information in central and peripheral locations; flat telecommunication rates.

— *Economy*. With its 450 million consumers and its multinational labour force the European Community presents an economic empire of unprecedented magnitude. However, economic growth is unevenly distributed in Europe. The central European countries have fared best, while the southern countries have to rely on less profitable tourist and service economies. Even in the richer countries income disparities have increased due to redistributional tax policies and declining labour union influence.

— *Environment*. The car and truck populations of Europe have doubled since 1990. So has congestion and pollution on motorways and in urban areas. A completely new network of European toll motorways is nearing completion, some of them cut through the last remaining national parks. Congestion in London is controlled by rigorous road pricing. Rome and Athens are drowned under a sea of illegally parked cars and permanent smog. Most of their antiquities are permanently damaged.

— *Regional development*. The Single European Market has resulted in a further spatial concentration of economic activities. The high-accessibility belt from London to Milan has become a veritable megalopolis with a population of eighty million which already in the 1990s, due to the booming east European economies, spread out its tentacles as far as Berlin and Vienna. The negative side effects are severe agglomeration diseconomies in its centres characterized by exploding land prices, congestion and environmental deterioration, while the peripheral regions suffer from economic decline and depopulation.

— *Urban and rural form*. The typical major city today consists of three distinct parts: the *international* city centre where incomes, rents, shops and restaurants follow standards set in Tokyo, London and New York; the *middle-class* suburbs where the majority of wage earners live a comfortable life; and the *underclass* ghettos were jobless people and illegal immigrants are supported by welfare. Manufacturing has moved out to industrial parks near motorway exits. The countryside close to agglomerations and in resort areas has practically disappeared.

— *Goods transport*. During the 1990s, the dramatic increase in road freight traffic caused congestion on motorways and unacceptable levels of noise and pollution in cities. After the turn of the century a gigantic building programme for an entirely new network of European toll motorways, bridges and tunnels financed by private capital improved road transport capacity to meet demand. Most truck freight is handled via strategically located freight centres by a small number of transnational carriers. Railway freight services have practically disappeared.

— *Passenger transport*. Never has there been such a mobile society. Fuelled by continued economic prosperity, individual auto-mobility dominates local and regional travel. Road transport informatics and automated driving have made the use of the automobile feasible for a larger proportion of the population. Local public transport has been reduced to a feeder service for the automobile. High-speed trains link most European cities with each other and compete with airlines operating from more and smaller airfields. The capacity of the large airports and the airspace over Europe has been reached.

— *Communications*. The growth of the European economy is built on information transmitted over high-capacity fibre optics and satellite telecommunications networks. The 'wired' or 'informational' city creates new life-styles and mobility patterns. It has become a status symbol of the information-rich, active and mobile to be informed, monitored, served and entertained by a host of 'intelligent' devices such as cellular telephones, pagers or voice-recognition hand-held computers. The telecommunications revolution has favoured the central European regions at the expense of the periphery.

However, there was also disagreement. Is this not a too naïve extrapolation of existing trends? The consequences of the unconstrained growth scenario are so frightening that it is hard to believe that there will be no controlling action by national and European governments. So a great number of modifications of the growth scenario were suggested, some of which were taken from the other two scenarios.

It was argued, for instance, that the polarizing effects of the Single European Market are not altogether clear, and any inequalities it creates would certainly be compensated by the Community. A surprisingly large number of respondents seemed to put much faith in the will and ability of the Community (or the European government) to counteract the growing spatial polarization in Europe. The outlook on the environment in the growth scenario also appeared to many as too grim. They pointed out that the advances in environmental protection which have already been made, will continue due to strong environmentalist movements and that through measures such as taxes on car ownership and petrol the environmental situation can be kept constant or even improved. Many also saw a brighter future for the European city pointing to the observable signs of back-to-the-city movements and inner-city restoration. The views on freight transport showed a large degree of uncertainty. To rely on road transport as in the growth scenario seemed untenable to many, but given the present state of rail freight services no clear alternative trend emerged. Similarly, many felt that a society totally built on the private car would not be feasible for passenger transport and that a mixed policy of promoting both the car and the train would be more likely. With respect to telecommunications, many felt that telecommunications would also contribute to more equity.

In summary, the modifications and suggestions tended to endorse the growth scenario but added some more moderate, less radical notes to it. Nevertheless the outlook remains rather gloomy. Even if the experts are only partly right, the most likely scenario for transport and communications in Europe is a veritable horror scenario. It presents a continent with an unprecedented level of material wealth and technological perfection yet with unparalleled spatial disparities between its regions and cities, congested roads and a collapsed public transport system, a disappearing countryside and a devastated environment. Are there no alternatives?

The most preferred scenario

As pointed out in Chapter 12, there was a great difference between the scenario the respondents thought to be the most realistic and the one they most preferred. To be true, there were some respondents for whom the most likely scenario was also the most preferred. They felt that growth is so essential for the functioning of market economies that they could not imagine any solution without growth as problems of social equity and environmental pollution could be successfully attacked only under growth. Undoubtedly there is some empirical evidence for the validity of this line of argument.

The majority of respondents, however, felt that a fundamental change in values and policy-making is required. Both the equity paradigm and the environment paradigm had their supporters, but clearly the environment paradigm turned out to be the winner.

In this section the scenario most preferred by the respondent group as a whole will be sketched out as a combination of the equity and environment component scenarios. The essential characteristics of this combined target scenario is summarized on the right-hand side of Table 13.1.

— *Population.* In 2010, the European social security system underwent a deep crisis. With the growing number of old people the demand for health and social services grew while the taxable working population decreased. The way out was to encourage immigration from developing countries with growing populations. However, the decline in population also reduced the pressure on environmental resources, on land prices in cities and on traffic congestion. Because the government supports young families, fertility rates have started to rise in Germany, Holland and Sweden.

— *Life-styles*. In the early 2000s, a fundamental change of values took place in Europe. Young people were alienated by the 'elbow mentality' of the older generation and wanted a world of equal opportunity and social peace built on solidarity instead of competition. The family, having children, to be at home with friends and other traditional values have experienced an unexpected renaissance; participation in community affairs, team building and consensus formation are emphasized. The government promotes three-generation families by tax incentives and housing support for young people.

— *Economy*. The European government is primarily concerned with promoting sustainable development. It has imposed taxes on luxury goods and services in order to reduce consumer demand and it has also raised production costs by improving health and safety standards for consumer products and at work places as well as enforcing rigorous emission standards for manufacturing and transport. A high proportion of the government R&D budget is now spent on the development of alternative technologies which are environmentally more acceptable and reduce the wastage of non-renewable resources.

— *Environment*. By 2020, Europe has become a leader in environment-conscious policy-making. Emission standards for industry and transport are stricter than in the USA and Japan, the use of fossil fuels has been kept constant since 1995. Heavy taxes on car ownership and petrol brought car ownership down and made public transport almost profitable. Many people, not always voluntarily, moved back into the city; this was good for small cities, but in large metropolitan areas commuting times have become excessive.

— *Regional development*. Decentralization programmes such as the *Technopolis Network*, the *Remote Area Highway Programme* and the *Regional Airport Scheme* were coupled with strict land-use control in urban areas, tax incentives for location in non-metropolitan regions and flat rates for long-distance telecommunication services. Rigorous policies to force industry and consumers to adopt environment-conscious behaviour had far-reaching consequences for the European regions. Transport in general has become cleaner, but also more expensive.

— *Urban and rural form*. The government pursued a policy mix of incentives and regulations to reduce agglomeration diseconomies at the top of the urban hierarchy and give financial aid for small rural cities. Through property gains taxes and value capture legislation it was possible to keep land speculation in the large cities within limits. Because car use had become extremely expensive, most people looked for a home close to their work. Public transport was made more efficient, and the resulting reduction in car traffic made it possible to downgrade some trunk roads or even convert them into parks.

— *Goods transport*. The 'green' solution has been to restrict heavy road traffic and to increase its cost. Freight operations by air have been limited by regulation and heavy taxes. Thus rail has kept a larger share of goods transport. Recently new goods aircraft propelled by liquid hydrogen and clean fuel engines for road vehicles have become available. These quiet and clean 'ecological vehicles' are used to support a more decentralized society. A dispersed network of low-density settlements is gradually replacing some of the environmentally ineffective 'city deserts'.

— *Passenger transport*. The decision to constrain the automobile was not popular, but the government convinced the public that the state of the environment made it necessary. A massive programme to revitalize public transport brought a renaissance of trolleybuses, trams, metros and electric trains. Cars operated with hydrogen from solar collectors on homes will eventually provide environmentally harmless mobility for a dispersed society.

— *Communications*. Telecommunication is now considered to be an important instrument to decrease differences between the industrial and governmental centres and the periphery. Via telephone, telefax and data lines firms in remote villages have on-line access to industrial data bases. All telecommunication charges are independent of the distance from the information source. This policy has contributed much to decreasing the inequalities between central and peripheral life.

Like the other scenarios in this book, this scenario cannot claim to be a rigorous and comprehensive picture of the state of transport and communications in Europe in the year 2020. Rather it is a collage or patchwork of metaphors and it must be left to the reader to fill it with life.

However, it may have become clear that the Europe 2020 which emerges from behind this collage is very different from the one behind the 'most likely' scenario. It describes a more resource-conserving and sustainable, but also more equitable and gentle world. The question is whether it can be achieved without severe losses of affluence and convenience. Some of the respondents did not believe that and therefore stayed with the growth scenario, hoping that, with the help of advanced technology, there is a corridor of benign growth. The others had no more confidence in technology.

Choices for decision makers

What conclusions can be drawn from this analysis that might be useful for decision-makers in Europe? The authors feel strongly that, despite the limited representativeness of the sample, the results suggest that European policy-makers are at a crossroads where two fundamentally opposed directions of political action can be chosen. In the most basic dimension, the choices are again determined by the three paradigms growth, equity and environment. Constituencies and decision-makers will have to choose whether they want to continue to follow the growth paradigm for the sake of being competitive in the world market against the United States and Japan, or whether they want to follow an alternative course.

The first direction is the one currently followed by national governments and the European Commission. Its basic paradigm is that because of global competition with the United States and Japan, Europe must do everything possible to modernize its infrastructure and manufacturing equipment and hence promote continuous growth. Underlying this philosophy is that only a growing European economy can pay for the large amount of investment necessary for this global race. However, planning for growth in one of the richest regions of the world means widening the gap between the industrialized and developing countries with unpredictable consequences for the future. In a competitive economy it also means condoning spatial polarization, because modernization is most efficient in the most advanced and most affluent metropolitan regions in the European core. Condoning spatial polarization, however, means accepting growing income disparities between the core regions and the regions at the European periphery, which undoubtedly will benefit from the growth of the centres, but inevitably will grow less than these. In addition, further spatial polarization means more agglomeration diseconomies in terms of congestion, land speculation and environmental damage.

The other direction would be to promote an ecologically sustainable and socially equitable future at the expense of economic growth. Such a strategy would not only work towards a more peaceful solution of the imminent conflict between the developing and the industrialized countries, but would also avoid many of the negative prospects of spatial polarization and environmental degradation which are inevitably associated with continued economic growth. It would be a great challenge for Europe to demonstrate that there is a future for Europe that is both equitable and in balance with nature without excessive and destructive material growth.

In more concrete terms, the choices for the nine component fields are, in the order of urgency indicated in Chapter 12, as follows (see Table 13.2).

CHOICES FOR EUROPE

Table 13.2 Summary of choices for action in order of urgency

Field	Growth	Equity/Environment
Regional development	Continue to let capital flow to already prosperous central regions; continue to concentrate transport and communication infrastructure in core regions.	Promote decentralized system of regions with autonomy to develop their endogenous potential; promote deconcentration of infrastructure.
Urban and rural form	Continue to promote winner cities by concentration of high-level infrastructure and not intervening in destructive competition.	Promote small and peripheral cities through modern infrastructure and government agencies; contain urban sprawl in agglomerations.
Goods transport	Continue to reward road freight transport through low taxation and motorway construction.	Reduce road freight transport through taxation and regulation; invest in combined road/rail transport; reduce volume of goods.
Passenger transport	Continue to promote car traffic through insufficient taxation, road construction, dispersed settlement of structures and neglect of railways.	Discourage use of cars by taxation and road pricing; substantially improve the attractiveness of public transport; promote reurbanization and mixed land use.
Life-styles	Continue to promote competition and egoism in education and economic life.	Support families to reverse decline in household size; promote shift in values from individual to collective goals.
Communications	Continue to promote concentration of telecommunications infrastructure in core regions.	Subsidize telecommunications in peripheral regions; introduce flat rate telecommunications charges.
Economy	Continue to serve needs of transnational companies; continue deregulation of economy; continue protectionism against developing countries.	Promote small and medium-sized companies; promote equalization of incomes and social security within Europe; promote economic cooperation with Africa.
Environment	Continue to settle for lowest common denominator for environmental standards; continue to promote high-speed transport infrastructure.	Plan for sustainable development; adopt environmental standards of most advanced countries; redirect transport investments to peripheral regions.

Population	Continue to permit non-permanent immigration of workers from non-EC countries without giving them full citizenship.	Permit controlled permanent immigration from non-EC countries; improve support for families.

Regional development

either to let capital and economic activity continue to move to the already prosperous central regions of Europe without adequately paying for the environmental damage and agglomeration diseconomies they inflict on themselves and others and to continue to promote the concentration of high-speed transport and high-level telecommunications infrastructure in the already overcrowded European agglomeration and air corridors;

or to promote a truly decentralized system of European regions with a high degree of regional autonomy enabling them to develop their endogenous potential and narrow the gap between rich and poor regions; to promote the deconcentration of agglomerations to reduce congestion and other agglomeration diseconomies; to promote the transformation of European transport networks towards a dispersed, medium-speed but high-efficiency network without peak loads in space or time.

Urban and rural form

either to continue to promote the 'winner' cities by concentrating high-level infrastructure, public facilities and subsidies in the largest cities, by rewarding wasteful and environmentally harmful modes of transport and by not intervening in the destructive competition of large and small cities against each other and in the exploitation of natural resources through urban sprawl through a *laissez-faire* regional planning policy;

or to engage in an active anti-agglomeration strategy by promoting peripheral small and medium-sized cities through modern transport and communications infrastructure, by decentralizing government agencies and by promoting local industries through seed funds and tax incentives while at the same time constraining further urban sprawl at the outskirts of large agglomerations through greenbelt policies or other development controls.

Goods transport

either to continue to reward road freight transport through low taxation of lorries, aggressive motorway construction, neglect of rail freight services outside large agglomerations and retarded introduction of stricter emission controls for trucks, and to promote further long-distance transport of products that can be produced and consumed regionally such as vegetables, flowers or beverages, and to tolerate and even subsidize the establishment of dispersed, space-consuming production facilities relying on transport-intensive just-in-time logistics;

or   to reduce road freight transport through fair taxation of lorries taking account of the damage of lorries to roads and the environmental costs following the 'polluter-pays' principle, to introduce immediately rigorous environmental standards for lorries and traffic restrictions for lorries in environmentally sensitive areas and congested areas in cities during rush hours or at night in residential districts, to invest substantially in combined road/rail transport facilities and services and to reduce the volume of freight by the relocation of heavy process industries, disincentives for excessive just-in-time logistics and the promotion of regional, short-distance distribution networks.

Passenger transport

either   to promote further increases of car traffic through insufficient taxation of car ownership and car use, slow introduction of more rigorous environmental standards for cars and unconstrained continuation of road construction even in countries with fully developed road networks, while at the same time reducing investment and operating subsidies to railways and public transport; and to promote the trend to dispersed, space-consuming and transport-intensive settlement structures, which cannot be economically served by public transport, without concurrent decentralization of employment;

or   to discourage the use of cars by a differentiated system of taxation, user fees and road pricing taking full account of the environmental and social costs of car usage in general and in particular areas or times of day, while at the same time substantially improving the attractiveness of public transport by new investments, enhanced service and competitive fare structures (acknowledging the fact that public transport is a public good that needs to be subsidized); and to improve the spatial interaction between residences, work places, shopping areas and public facilities by promoting reurbanization and mixed land use.

Life-styles

either   to accept the highly individualistic and resource-consuming society that is depicted in the growth scenario, together with its dark side of poverty and increasing disparities within society;

or   to promote some combination of the equity and environmental scenarios and return to a collective and caring society accompanied by a renaissance of the family, albeit in forms very different from our Victorian forebears to reverse the decline in household size.

Communications

either   to continue to promote the demand-driven concentration of high-level telecommunications infrastructure in core regions—the *laissez-faire* option with industry and business as leaders of telecommunications development—and to proceed with the deregulation of telecommunication services resulting in insufficient services for businesses in peripheral low-demand regions.

or   to promote telecommunications in peripheral regions by subsidizing investments for telecommunications infrastructure and introducing flat rate (not distance-dependent) telecommunication charges—this is the decentralization option available to dampen the growth of the largest agglomerations by establishing equivalent service, culture and information work opportunities in small cities and peripheral regions.

Economy

either   to accept the main features of the growth scenario with its unparalleled wealth and equally unparalleled congestion and consumption of resources. Such a scenario would lead to greater regional disparities within Europe;

or   to develop some combination of the equity and environment scenarios which seek to reduce regional disparities and encourage sustainable development. Such a scenario would involve stringent measures on resource-consuming activities which would reduce Europe's overall competitiveness in relation to growth-driven economies in other parts of the world.

Environment

either   to continue to pay lip-service to ecological concerns but also to make concessions to growth interests of countries and powerful industry groups; to settle down with the lowest common denominator between the European countries where environmental standards are concerned; and to continue to promote one-sidedly high-speed transport infrastructure for the sake of global economic competitiveness and efficiency;

or   to join the few enlightened industry leaders who realize that for environmental and global equity reasons the aggressive growth of the industrialized countries must be slowed down; to orient European environmental policy not at the level of environmental protection of the least but the most advanced nations; to redirect a substantial share of transport investment to improving environmentally desirable transport modes and transport infrastructure in peripheral regions.

Population.

either   to continue to permit non-permanent immigration of workers from non-EC countries without giving them full citizen status as suggested in the growth scenario. Such a policy would relieve the shortage of labour due to population decline in some European countries but would permanently establish a society based on discrimination;

or   to accept that Europe will become an immigration continent and to permit controlled permanent immigration from non-EC countries with growing populations as suggested in the equity scenario. Combined with a policy to support families and working parents as in the environment scenario, this would help[to offset the ageing of the population and produce a more balanced age structure and labour market and would help to reduce the gap between the rich and poor countries.

Critics will argue that the world is not black or white and that the choices are not as simple. This is certainly true, and the authors realize that the real path to Europe 2020 will be a long, and sometimes painful, history of conflicts, negotiations and compromises between and within countries and regions.

However, although the details may be open to debate, there can be no doubt that the *principal* decision to be made is fundamental and concerns our attitude to growth. If we, the industrialized countries, insist on continuing to base our culture on the self-propelling dynamics of unconstrained economic growth, we will not only create a world which despite its wealth may not be worth living in, but will also provoke a conflict of still unknown dimensions with the deprived majority of the globe. To believe that only our economic growth will enable us to help the developing world is a dangerous, and politically naïve, illusion. What we have to learn is that a sustainable world will not be achieved by sharing our wealth with them but by sharing a part of their poverty.

The remaining question is whether there is a future for Europe that is both equitable and in balance with nature which avoids the pitfalls of unharnessed growth. Certainly it will not be a future without material growth as much of our wealth is based on competition in the market. The great challenge for shaping the geography of Europe's future lies in the exploration of solutions in which controlled growth is guided in a socially balanced and environmentally sustainable way.

REFERENCES

Albrechts, L., Moulaert, F., Roberts, P. and Swyngedouw, E. (eds), 1989, *Regional policy at the cross-roads*, Jessica Kingsley Publishers, London.

Amin, A. and Goddard, J.B. (eds), 1986, *Technological change, industrial restructuring and regional development*, Allen and Unwin, London.

Aydalot, P. and Keeble, D., 1988, *High technology industry and innovative environments: the European experience*, Routledge and Kegan Paul, London.

Badcock, B., 1984, *Unfairly structured cities*, Blackwell, Oxford.

Banister, D. (ed.), 1989, 'The final gridlock', *Built Environment*, **3/4**.

Barde, J.P. and Pearce, D.W. (eds), 1990, *Valuing the environment and decision-making*, Earthscan Publications, London.

Barney, G.O. (ed.), 1980, *The Global 2000 report to the President: entering the twenty-first century*, US Government Printing Office, Washington DC.

Bell, D., 1973, *The coming of post-industrial society: a venture in social forecasting*, Basic Books, New York.

Bieber, A. and Potier, F., (forthcoming), 'Transport and the development of tourism: some European scenarios' in G. A. Giannopoulos, A. Gillespie and S. Wandel (eds), *Transport, communications and spatial organization in the Europe of the future*.

Blackburn, P., Coombs, R. and Green, K., 1985, *Technology, economic growth and the labour process*, MacMillan, New York.

Bourgeois-Pichat, J., 1981, 'Recent demographic change in western Europe: an assessment', *Population and Development Review*, **7**:19-42.

Bourgeois-Pichat, J., 1984, 'Mortality trends in the industrialised countries: mortality and health policy', ST/ESA/Ser A/91, United Nations, New York.

Brotchie, J., Batty, M. and Hall, P. (eds), 1990, *New technologies and spatial systems*, Unwin Hyman, London.

Brotchie, J., Hall, P. and Newton, P.W. (eds), 1987, *The spatial impact of technological change*, Croom Helm, London.

Brotchie, J., Newton, P., Hall, P. and Nijkamp, P. (eds), 1985, *The future of urban form: the impact of new technology*, Croom Helm, London.

Buchanan, C., 1963, *Traffic in towns: a study of long-term problems of traffic in urban areas*, Ministry of Transport, HMSO, London.

Button, K.J., 1982, *Transport economics*, MacMillan, London.

Button, K.J. and Banister, D. (eds), 1991, *Transport in a free market economy*, MacMillan, London.

Button, K.J. and Barde, J.-P. (eds), 1990, *Transport policy and the environment: six case studies*, Earthscan Publications, London.

REFERENCES

Button, K.J. and Gillingwater, D., 1986, *Future transport policy*, Routledge, London.

Castells, M., 1977, *The urban question: a Marxist approach*, Edward Arnold, London.

Castells, M., 1989, *The informational city: information technology, economic restructuring, and the urban-regional process*, Basil Blackwell, Oxford.

Cecchini, P., 1988, *The European challenge: the benefits of a single market*, Wildwood House, Aldershot.

Champion, A.G. (ed.), 1989, *Counter-urbanization: the changing pace and nature of population deconcentration*, Edward Arnold, London.

Cheshire, P.C. and Hay, D.G., 1989, *Urban problems in western Europe: an economic analysis*, Unwin Hyman, London.

Close, P. and Collins, R. (eds), 1985, *Family and economy in modern society*, MacMilian, London.

Coale, A. and Watkins, S.C., 1986, *The decline of fertility in Europe*, Princeton University Press, Princeton, NJ.

Commission of the European Communities, 1986, *The state of the environment in the European Community 1986*, Office for Official Publications of the EC, Luxembourg.

Commission of the European Communities, 1990, *Green Paper on the urban environment*, Office for Official Publications of the EC, Luxembourg.

Commission of the European Communities, 1991a, *The regions in the 1990s: fourth periodic report on the social and economic situation and development of the regions of the Community*, Office for Official Publications of the EC, Luxembourg.

Commission of the European Communities, 1991b, *Europe 2000: outlook for the development of the Community's territory*, Office for Official Publications of the EC, Luxembourg.

Cooke, P. (ed.), 1989, *Localities: the changing face of urban Britain*, Unwin Hyman, London.

Cooper, J., Browne, M. and Peters, M., 1991, *European logistics-markets, management and strategy*, Blackwell Publishers, Oxford.

Council of Europe, 1990, *Recent demographic developments in the member states of the Council of Europe*, Council of Europe, Strasbourg.

Crick, B. (ed.), 1984, *George Orwell: 1984*, Clarendon Press, Oxford.

Crouch, C. and Marquand, D. (eds), 1990, *The politics of 1992: beyond the Single European Market*, Basil Blackwell, Oxford.

Davis, K., Bernstam, M.S. and Ricardo Campbell, R. (eds), 1986, 'Below replacement fertility in industrialised countries', *Population and Development Review*, **12**, supplement.

de Jong, H.W. (ed.), 1988, *The structure of European industry*, Kluwer, Dordrecht.

Enderwick, P. (ed.), 1989, *Multinational service firms*, Routledge, London.

Euro Disney S.C.A., 1990, *Annual report*, Euro Disneyland, Villiers sur Marne.

European Conference of Ministers of Transport, 1989a, 'Telematics in goods transport', Report of the 78th Round Table on Transport Economics, OECD Publication Services, Paris.

European Conference of Ministers of Transport, 1989b, *Rail network co-operation in the age of information technology and high speed*, OECD Publications Service, Paris.

European Round Table of Industrialists, 1988, *Keeping Europe mobile*, European Round Table of Industrialists, Brussels.

European Round Table of Industrialists, 1990, *Logistics and transport in Europe*, European Round Table of Industrialists, Brussels.

European Round Table of Industrialists, 1991, *Missing networks: a European challenge: proposals for the renewal of Europe's infrastructure*, European Round

REFERENCES

Table of Industrialists, Brussels.

Eurostat, 1988/1989/1990, *Regions*, statistical yearbook 1987/1988/1989, Office for Official Publications of the EC, Luxembourg.

Eurostat, 1988/1989/1990, *Basic statistics of the Community*, 25/26/27th edn, Office for Official Publications of the EC, Luxembourg.

Ewers, H. J. and Allesch, J. (eds), 1990, *Innovation and regional development: strategies, instruments and policy coordination*, Walter de Gruyter, Berlin.

Faulks, R.W., 1990, *Principles of transport*, 4th edn, McGraw Hill, New York.

Findlay, A. and White, P. (eds), 1986, *Western European population change*, Croom Helm, London.

Fowles, J. (ed.), 1978, *Handbook of futures research*, Greenwood Press, Westport, CT.

Freeman, C. and Jahoda, M. (eds), 1978, *World futures: the great debate*, Martin Robertson, London.

Gershuny, J., 1978, *After industrial society? The emerging self-service economy*, Macmillan, London.

Gershuny, J. and Miles, I., 1983, *The new service economy: the transformation of employment in industrial societies*, Frances Pinter, London.

Giannopoulos, G.A., Gillespie, A. and Wandel, S. (eds), forthcoming, *Transport, communications and spatial organization in the Europe of the future*.

Giaoutzi, M. and Nijkamp, P. (eds), 1988, *Informatics and regional development*, Avebury, Aldershot.

Giaoutzi, M., Nijkamp, P. and Storey, D.J. (eds), 1988, *Small and medium-size enterprises and regional development*, Routledge, London.

Goddard, J. and Gillespie, A. (eds), 1988, 'Advanced telecommunications and regional economic development', in M. Giaoutzi and P. Nijkamp (eds), 1988, *Informatics and regional development*, Avebury, Aldershot, pp. 121–46.

Group Transport 2000 Plus, 1990, *Transport in a fast changing Europe: Vers un reseau Européen des systèmes de transport*, Group Transport 2000 Plus, Brussels.

Hägerstrand, T., 1987, 'Human interaction and spatial mobility: retrospect and prospect', in P. Nijkamp and S. Reichman (eds), 1987, *Transportation planning in a changing world*, Gower, Aldershot, pp. 11–28.

Hall, P., 1963, *London 2000*, Faber, London.

Hall, P. (ed.), 1976, *Europe 2000*, Duckworth, London.

Hall, P., 1984, *The world cities*, 3rd edn., Weidenfeld and Nicolson, London.

Hall, P., 1987, 'The geography of high technology: an Anglo-American comparison', in J. Brotchie, P. Hall, and P.W. Newton, (eds), 1987, *The spatial impact of technological change*, Croom Helm, London, pp. 11–12.

Hall, P., 1988, *Cities of tomorrow: an intellectual history of urban planning and design in the twentieth century*, Basil Blackwell, Oxford.

Hall, P., 1989, *London 2001*, Unwin Hyman, London.

Hall, P. and Hass-Klau, C., 1985, *Can rail save the city? The impacts of rail rapid transit and pedestrianisation on British and German cities*, Gower, Aldershot.

Hall, P. and Hay, D., 1980, *Growth centres in the European urban system*, Heinemann Educational Books, London.

Hall, P. and Markusen, A. (eds), 1985, *Silicon landscapes*, Allen and Unwin, Boston.

Hall, R., 1988, 'Recent patterns and trends in European households at national and regional scales', *Espaces, Populations, Sociétiés*, **1988/1**:13–32.

Hamnett, C., McDowell, L. and Sarre, P. (eds), 1989, *The changing social structure*, Sage, London.

Handy, C., 1989, *The age of unreason*, Business Books, London.

Harvey, D., 1973, *Social justice and the city*, Edward Arnold, London.

Hass-Klau, C., 1990, *The pedestrian and city traffic*, Belhaven Press, London.

Hepworth, M., 1989, *The geography of the information economy*, Belhaven Press, London.

Hirschfield, A., 1991, *Urban deprivation: problems, process and counter-measures in developed countries*, Belhaven Press, London.

Hirschhorn, L., 1980, 'Scenario writing: a developmental approach', *Journal of the American Institute of Planners*, **46**:172–83.

Hoem, J.M., 1990, 'Social policy and recent fertility change in Sweden', *Population and Development Review*, **16**: 735–49.

IIASA, 1990, 'CO_2: a balancing of accounts', *Options*, December 1990, pp. 10–13.

Jacobs, J. (1962), *The death and life of great American cities*, Jonathan Cape, London.

Jansen, G.R.M., Nijkamp, P. and Ruijgrok, C. (eds), 1985, *Transportation and mobility in an era of transition*, North Holland Publishing Co., Amsterdam.

Japan Institute for Social and Economic Affairs, 1989, *Japan 1990: an international comparison*, Japan Institute of Social and Economic Affairs, Tokyo.

Johnson, S.P. and Corcelle, G., 1989, *The environmental policy of the European Communities*, Graham & Trotman, London.

Kahn, H. and Wiener, A.J., 1967, *The year 2000: a framework for speculation on the next thirty-three years*, Macmillan, New York.

Keeble, D. and Wever, E. (eds), 1986, *New firms and regional development in Europe*, Croom Helm, London.

Keeble, D., Offord, J. and Walker, S., 1986, 'Peripheral regions in a community of twelve member states', Final report for the Commission of the European Communities, Department of Geography, University of Cambridge, Cambridge.

Keyfitz, N., 1989, *On future mortality*, WP 89–59 International Institute for Applied Systems Analysis, Laxenburg, Austria.

King, A. and Schneider, B., 1991, *The first global revolution: report of the Club of Rome*, Pantheon Books, New York.

King, R., 1990, 'The social and economic geography of labour migration: from guestworkers to immigrants' in D. Pinder, (ed.), *Western Europe; challenge and change*, Belhaven Press, London.

Kinsman, F., 1987, *The telecommuters*, John Wiley, London.

Kunzmann, K.R., 1989, 'Ausflug ins Morgenland' (Excursion to Tomorrowland), *Geo Special*, **3**: 32–8.

Kunzmann, K.R. and Wegener, M., 1991, *The pattern of urbanization in Western Europe 1960–1990*, Report 28, Institute of Spatial Planning, University of Dortmund, Dortmund.

Läpple, D., 1990, 'Vom Gütertransport zur logistischen Kette' (From goods transport to logistic chain), *Mitteilungen der Deutschen Akademie für Städtebau und Landesplanung*, **34** (1): 11–33.

Leridon, H., 1990a, 'Cohabitation, marriage separation: an analysis of life histories of French cohorts from 1968 to 1985', *Population Studies*, **44**: 127–44.

Leridon, H., 1990b, 'Extra marital cohabitation and fertility', *Population Studies*, **44**: 469–88.

Lundberg, D.E., 1990, *The tourist business*, 6th edn., Van Nostrand Reinhold, New York.

Lutz, W. (ed.), 1991, *Future demographic trends in Europe and North America: what can we assume today?*, Academic Press, London.

McKinnon, A., 1989, *Physical distribution systems*, Routledge, London.

Mackintosh, I., 1986, *Sunrise Europe: the dynamics of information technology*, Blackwell, Oxford.

Martin, P.L., Honnekop, E. and Ullmann, H., 1990, 'Europe 1992: effects on labour migration', *International Migration Review*, **24**: 591–603.

Mason, R., Jennings L. and Evans R., 1983, *Xanadu: the computerized house of tomorrow and how it can be yours today*. Acropolis Books, Washington DC.

Mason, R., Jennings, L. and Evans, R., 1984, 'A day at Xanadu: family life in tomorrow's computerized house', *The Futurist*, February, pp. 17–24.

Masser, I. and Foley, P., 1987, 'Delphi revisited: expert opinion in urban analysis', *Urban Studies*, **24**: 217–25.

Masuda, Y., 1981, *The Information Society as post industrial society*, Institute for the Information Society, Tokyo.

Meadows, D.H., Meadows, D.L., Randers, L. and Behrens, W.W., 1972, *The limits to growth: a report for the Club of Rome's project on the predicament of mankind*, Earth Island, London.

Miles, I., 1988, *Home informatics*, Frances Pinter, London.

Miles, I., Rush, H., Turner, K. and Bessant, J., 1988, *Information horizons: the long-term social implications of the new information technologies*, Edward Elgar, Aldershot.

Ministry of Transport and Public Works, 1989, *Second transport structure plan*, Ministry of Transport and Public Works, Den Haag.

Mogridge, M.J.H., 1990, *Traffic in towns: jam yesterday, jam today and jam tomorrow?*, Macmillan, London.

Monck, C.S.P., Porter, R.B., Quintas, R.R., Storey, D.J. and Wynaczyk, P., 1988, *Science parks and the growth of high technology firms*, Croom Helm, London.

Moss, M., 1987, 'Telecommunications and international financial centres', in J. Brotchie, P., Hall, and P.W., Newton, (eds), 1987, *The spatial impact of technological change*, Croom Helm, London, pp. 75–88.

Mumford, L., 1961, *The city in history: its origins, its transformations and its prospects*, Harcourt, Brace, New York.

Munton, K.G., 1986, 'Past and present future life expectancy increases at later ages: their implications for the linkage of chronic morbidity, disability and mortality', *Journal of Gerontology*, **41**: 672–81.

Myers, N. (ed.), 1990, *The atlas of future worlds*, Robertson McCara 9, London.

Naisbitt, J., 1984, *Megatrends: ten new directions transforming our lives*, Macdonald, London.

Naisbitt, J. and Aburdene, P., 1990, *Megatrends 2000: the next ten years . . . major changes in your life and world*, Pan Books, London.

Nijkamp, P. (ed.), 1986, *Technological change, employment and spatial dynamics*, Springer Verlag, Berlin.

Nijkamp, P. (ed.), 1990, *Sustainability of urban systems: a cross-national evolutionary analysis of urban innovation*, Avebury, Aldershot.

Nijkamp, P. and Reichman, S. (eds), 1987, *Transportation planning in a changing world*, Gower, Aldershot.

Nijkamp, P., Reichman, S. and Wegener, M. (eds), 1990, *Euromobile: transport, communications and mobility in Europe*, Avebury, Aldershot.

Northcott, J. *et al.*, 1991, *Britain in 2010: the PSI report*, Policy Studies Institute, London.

OECD, 1986a, *Tourist policy and international tourism in OECD countries*, OECD Publications Services, Paris.

OECD, 1986b, *Fighting noise: strengthening noise abatement policies*, OECD Publications Services, Paris.

OECD, 1986c, *Urban policies in Japan*, OECD Publications Services, Paris.

OECD, 1988a, *Transport and the environment*, OECD Publications Services, Paris.

OECD, 1988b, *Cities and transport: Athens, Gothenburg, Hong Kong, London, Los*

REFERENCES

Angeles, Munich, New York, Osaka, Paris, Singapore, OECD Publications Service, Paris.

OECD, 1989, *Telecommunication network-based services: policy implications*, OECD Publications Services, Paris.

Orishimo, I., Hewings, G. and Nijkamp, P. (eds), 1988, *Information technology: social and spatial perspectives*, Springer Verlag, Berlin.

Owens, S. and Owens, P.L., 1991, *Environment, resources and conservation*, Cambridge University Press, Cambridge.

Pahl, R.E. (ed.), 1988, *On work*, Basil Blackwell, Oxford.

Pearce, D.W. *et al.*, 1989, *Blueprint for a green economy*, Earthscan Publications, London.

Pinder, D. (ed.), 1990, *Western Europe: challenge and change*, Belhaven Press, London.

Prognos, 1989, 'The development of EC labour markets up to 2000', Prognos, Basel.

Quinet, E., 1990, 'The social cost of land transport', Environmental Monographs 32, OECD Publications Services, Paris.

RECLUS, 1989, 'Les villes "Européennes"', Rapport pour la DATAR, La Documentation Française, Paris.

Redmond, J. (ed.), 1970, *William Morris: News from nowhere or an epoch of rest*, Routledge and Kegan Paul, London.

Rees, J., 1990, *Natural resources: allocation, economies and policy*, Routledge, London.

Robins, K. and Hepworth, M., 1988, 'Electronic spaces, new technologies and the future of cities', *Futures*, **20**: 155–76.

Robinson, G.M., 1990, *Conflict and change in the countryside: rural society, economy and planning in the developed world*, Belhaven Press, London.

Rodwin, L. and Sazanami, H., 1991, *Industrial change and regional economic transformation: the experience of Western Europe*, Unwin Hyman, London.

Rogers, A. and Serow, W., (eds), 1988, *Elderly migration: an international comparative study*, Institute of Behavioural Sciences, University of Colorado, Boulder, CO.

Ruijgrok, C., Jansen, B., McKinnon, H. and Wandel, S. (eds), forthcoming, *Logistics platforms and city logistics*.

Salomon, I., 1988, 'Transportation-telecommunication relationships and regional development, in M., Giaoutzi, and P., Nijkamp, 1988, *Informatics and regional development*, Avebury, Aldershot, pp. 90–102.

Scott, A.J. and Storper, M. (eds), 1986, *Production, work, territory: the geographical anatomy of industrial capitalism*, Allen and Unwin, London.

Segal Quince Wicksteed, 1985, *The Cambridge phenomenon: the growth of high technology industries in a university town*, Segal Quince Wicksteed, Cambridge.

Shirley, S., 1987, 'The distributed office', *Journal of the Royal Society of Arts*, **135**: 503–9.

Smilor, R.W., Kozmetsky, G. and Gibson, D.V. (eds), 1988, *Creating the technopolis: linking technology, commercialisation and economic development*, Ballinger, Cambridge, MA.

Smith, N. and Williams, P. (eds), 1986, *Gentrification of the city*, Allen Unwin, Hemel Hempstead.

Snickars, F., 1987, 'The transportation sector in the communications society: some analytical observations', in P. Nijkamp, and S. Reichman, (eds), 1987, *Transportation planning in a changing world*, Gower, Aldershot, pp. 226–42.

Steinle, W., 1988, 'Telematics and regional development in Europe: theoretical considerations and empirical evidence', in M. Giaoutzi, and P. Nijkamp, 1988,

Informatics and regional development, Avebury, Aldershot, pp. 72–89.

Steinmann, G., 1991, 'Immigration as a remedy for the birth dearth: the case of West Germany', in W. Lutz, (ed.), 1991, *Future demographic trends in Europe and North America: what can we assume today?*, Academic Press, London.

Stöhr, W.B. (ed.), 1990, *Global challenge and local response: initiatives for economic regeneration in contemporary Europe*, Mansell, London.

Surtz, E. (ed.), 1964, *St. Thomas More: Utopia*, Yale University Press, Newhaven, CT.

Sviдén, O., 1983, 'Automobile usage in a future information society', *Futures*, **15** (6), December 1983, 478–90.

Sviдén, O., 1988, 'Future information systems for road transport: a Delphi panel derived scenario', *Technological Forecasting and Social Change*, **33**: 159–78.

Thwaites, A.T. and Oakey, R. P., 1985, *The regional economic impact of technological change*, Frances Pinter, London.

TNO, 1990, 'Verkeer en Vervoer in 2015' (Traffic and Transport in 2015), Institute for Spatial Organisation (INRO), TNO, Delft.

Toffler, A., 1970, *Future shock*, Bodley Head, London.

Toffler, A., 1980, *The third wave*, Random House, New York.

Tolley, R., (ed.), 1990, *The greening of urban transport: planning for walking and cycling in western cities*, Belhaven Press, London.

Trinity College, 1990, *Cambridge Science Park*, Trinity College, Cambridge.

Umweltbundesamt, 1990, *Daten zur Umwelt 1988/89* (Environmental Data 1988/89), Umweltbundesamt, Berlin.

Ungerer, H. and Costello, N.P., 1988, *Telecommunications in Europe: free choice for the user in Europe's 1992 market*, Commission of the European Communities, Brussels/Luxembourg.

van den Berg, L., Drewett, R., Klaassen, L.H., Rossi, A. and Vijverberg, C.H.T., 1982, *Urban Europe: a study of growth and decline*, Pergamon Press, Oxford.

Vanhove, N. and Klaassen, L.H., 1987, *Regional policy: a European approach*, Avebury, Aldershot.

Vester, F., 1990, *Ausfahrt Zukunft* (Exit to the future), Wilhelm Heyne Verlag, Munich.

Vianello, M. and Siemienska, F. (eds), 1990, *Gender and inequality: a comparative study of discrimination and participation*, Sage, London.

Vickerman, R.W., 1984, *Urban economics*, Philip Allan, London.

Wall, R., 1989, 'Leaving home and going it alone: a historical perspective', *Population Studies*, **43**: 369–89.

Wandel, S. and Ruijgrok, C., forthcoming, 'The freight sector response to changing spatial organization of production' in G.A. Giannopoulos, A. Gillespie and S. Wandel (eds) *Transport, communications and spatial organisation in the Europe of the future*.

Wattenberg, B.J. 1989, *The birth dearth is what happens when people in free countries don't have enough babies*, Phatos Books, New York.

Werner, B., 1986, 'Fertility trends in the UK and in thirteen other developed countries 1966–1986', *Population Trends*, **51**: 18–29.

White, P. and van der Knaap, G.A. (eds), 1985, *Contemporary studies in migration*, Geobooks, Norwich.

Whitelegg, J., 1988, *Transport policy in the EEC*, Routledge, London.

Whitelegg, J. and Holzapfel, H., 1991, *Transport for a sustainable future: the case for Europe*, Belhaven Press, London.

Williams, A. (ed.), 1987, *The West European economy*, Hutchinson Education, London.

Williams, A.M. and Shaw, G. (eds), 1988, *Tourism and economic development: West European experiences*, Belhaven Press, London.

World Bank, Bulatao, R.A., Bos, E., Stephens, P.W. and Vu, M.T. (eds), 1990, *World population projections 1989–1990: short- and long-term estimates*, John Hopkins University Press, Baltimore and London.

World Commission on Environment and Development (Brundtland Committee), 1987, *Our common future*, Oxford University Press, Oxford.

LIST OF RESPONDENTS

Austria

Manfred Fischer, Vienna University of Economics
Klaus Schuch, Vienna University of Economics

Denmark

Uffe Jacobsen, School of Economics, Copenhagen

Finland

Asa Enberg, Helsinki University of Technology
Heikki Eskelinen, University of Joensuu

France

Alain Bieber, INRETS, Arcueil
Michel Frybourg, CGPC, ICEE, Paris
Marie-Hélène Massot, INRETS, Arcueil
Bernard Gerardin, INRETS, Arcueil
Jean-Pierre Orfeuil, INRETS, Arcueil
Francis Papon, INRETS, Arcueil

Germany

Heinz-Jürgen Bremm, University of Dortmund
Klaus R. Kunzmann, University of Dortmund
Seungil Lee, University of Dortmund
Klaus Spiekermann, University of Dortmund
Simone Strähle, University of Dortmund
Anna Wegener, University of Hamburg
Michael Wegener, University of Dortmund

Greece

Georgios Giannopoulos, University of Thessaloniki
John Toskas, Transport Planner, Kalamaria

Ireland

Sean Barrett, Trinity College, Dublin

Israel

Eran Feilelson, Hebrew University of Jerusalem
Shalom Reichman, Hebrew University of Jerusalem
Eliahu Stern, Ben Gurion University of the Negev

Italy

Roberta Capello, Università Luigi Bocconi, Milan

Netherlands

Sten Bexelius, Rijkswaterstaat, Rotterdam
Piet H.L. Bovy, Rijkswaterstaat, Rotterdam
Gijsbertus R.M. Jansen, TNO, Delft
Jeroen Klooster, Rijkswaterstaat, Rotterdam
Peter Nijkamp, Free University of Amsterdam
Piet Rietveld, Free University of Amsterdam
Cees J. Ruijgrok, TNO, Delft
Leo P. Schippers, Free University of Amsterdam
Elly van Doorn-Klein, Rijkswaterstaat, Rotterdam
Jaap M. Vleugel, Free University of Amsterdam

Norway

Oystein Engebretsen, Institute of Transport Economics, Oslo

Portugal

João F. dos Reis Simões, Instituto Superior Técnico, Lisbon

Slovenia

Slavko Hanzel, Prometni Institut, Ljubljana

Spain

Ginés de Rus Mendoza, University of Las Palmas de Gran Canaria

LIST OF RESPONDENTS

Sweden

Mats Abrahamsson, Linköping Institute of Technology
Bertil Agren, Linköping Institute of Technology
Dan Andersson, Linköping Institute of Technology
Christina Dalberg, Linköping Institute of Technology
Lars Lundqvist, Royal Institute of Technology, Stockholm
Ove Svidén, Sten Wandels department at Linköping Institute of Technology
Sten Wandel, Linköping Institute of Technology
Kerstin Westin, University of Umeå

Switzerland

Rico Maggi, University of Zurich
Claudia Nielsen, University of Zurich

Turkey

Cetin Turgai Günal, Planner, Ankara
Gökhan Mentes, Metroplan, Ankara
Ali Türel, Middle East Technical University, Ankara

United Kingdom

Kenneth Button, Loughborough University
Heather Campbell, University of Sheffield
Arman Farahmand-Razavi, University of Sheffield
Anthony D.J. Flowerdew, Marcial Echenique & Partners, Cambridge
Andrew Gillespie, University of Newcastle-on-Tyne
Peter M. Jones, Oxford University
Ian Masser, University of Sheffield
David Simmonds, David Simmonds Consultancy, Cambridge

SUBJECT INDEX

SUBJECT INDEX

SUBJECT INDEX

AUTHOR INDEX

AUTHOR INDEX

AUTHOR INDEX

Toffler A. 22, 214
Tolley R. 90, 160, 214
Trinity College 60, 214
Turner K. 54, 212

Ullmann H. 211
Umweltbundesamt 72, 214
Ungerer H. 163, 175, 214

Vanhove N. 108, 214
Vester F. 90, 160, 214
Vianello M. 54, 214
Vickerman R.W. 126, 214
Vijverberg C.H.T. 127, 214
Vu M.T. 215

Walker S. 211

Wall R. 54, 214
Wandel S. xiii, 143, 208, 210, 213, 214
Watkins S.C. 38, 209
Wattenberg B.J. 38, 214
Wegener M. 127, 143, 159, 211
Werner B. 38, 214
Wever E. 69, 211
White P. 38, 210, 214
Whitelegg J. 144, 160, 214
Wiener A.J. 22, 211
Williams A.M. 54, 215
Williams A. 68, 214
Williams P. 127, 213
World Bank 30, 215
World Commission on Environment and Development 90, 215